Practice-Based Professional Development *for* Teachers *of* Mathematics

Margaret Schwan Smith

NATIONAL COUNCIL OF TEACHERS OF MATHEMATICS
Reston, Virginia

Copyright © 2001 by
THE NATIONAL COUNCIL OF TEACHERS OF MATHEMATICS, INC.
1906 Association Drive, Reston, VA 20191-9988
(703) 620-9840; (800) 235-7566; www.nctm.org
All rights reserved

ISBN 0-87353-504-9

Printed in the United States of America

In memory of
Susan Loucks-Horsley

TABLE OF CONTENTS

The Commission on the Future of the Standards recognized that the changes in content and practice envisioned by *Principles and Standards for School Mathematics* (NCTM 2000) would require dramatic changes in the way in which professional development was conceptualized and conducted. The commission asked the Professional Development and Status Advisory Committee (PDSAC) of the National Council of Teachers of Mathematics (NCTM) to develop a plan that would address the professional development needs of the NCTM membership. One component of the committee's proposed plan to support mathematics teachers' implementation of standards-based instruction was to commission a guide to effective professional development practice that would help teacher educators, teacher-leaders, staff developers, and supervisors design, conduct, and evaluate professional learning opportunities for teachers that could transform their knowledge, beliefs, and habits of practice. This book is the product of their vision.

I would like to thank the members of the PDSAC who provided assistance and support during the two-year period in which the book was written: Glenn Allinger, Douglas Owens, Melva Greene, Marilyn Hala (Staff Liaison), Ellen Hook, Steven Leinwand (Board Liaison 1998–2000), Brent McClain, Lew Romagnano (Chair 1997–1998), Pamela Schram (Chair 2001–2002), Douglas Smeltz (Chair 1998–1999), Clara Tolbert, Bert Waits (Board Liaison 2000–2002), and Stephanie Williamson. Their commitment to this project and to the author, as well as their input and feedback through every step of the process, were invaluable.

This book draws heavily on the work in which I have been engaged for the past decade. In particular, throughout the book, the work of the QUASAR (Quantitative Understanding: Amplifying Student Achievement and Reasoning) project provides a context for considering the challenges and triumphs that teachers face as they endeavor to change the way mathematics is taught and learned. The QUASAR project's work also provides a source of empirical data that supports many of the claims that are made here. The work of the COMET (Cases of Mathematics Instruction to Enhance Teaching) project, funded by the National Science Foundation (NSF), is the source of many of the insights on the potential of practice-based materials to support teacher learning that are discussed in the book. Some of the materials that are used in examples also came from the COMET project.

Edward Silver and Mary Kay Stein, extraordinary colleagues and collaborators, have deeply influenced my thinking about professional development. Many of the ideas presented here are based on the discussions we have had about our joint work on the QUASAR and COMET projects and on the presentations and articles that we have done (in various combinations) over the past ten years.

I would also like to acknowledge the individuals who provided specific assistance to this effort: Iris Weiss and the TE-MAT (Teacher Education Materials) project for providing reviews of professional development materials that served as the basis of the brief descriptions that appear in Appendix B; Melissa Boston, a doctoral student at the University of Pittsburgh and member of the COMET project team, for producing the reviews that appear in Appendix B; and Melissa Boston, Mark Driscoll, Mary Lindquist, Marilyn Mays, and Edward Silver for offering feedback on earlier versions of this manuscript.

More than a decade has passed since the National Council of Teachers of Mathematics (1989) introduced a set of standards for grades K–12 mathematics and forever changed our vision of school mathematics. The underlying goal of mathematics education espoused by this document is to create mathematically powerful students who can communicate with their teacher and their peers about the mathematics they are learning, who can argue convincingly and provide mathematical justifications to support their positions, and who can work alone and with peers to solve problems that require complex thinking and reasoning strategies.

The view of mathematics learning presented in this document stands in sharp contrast to the more traditional view, which features memorizing rules, manipulating symbols, and using algorithms to solve routine problems. In traditional practice, classrooms are more likely to follow the decades-old format in which teachers correct homework, lecture on a two-page textbook lesson, and supervise the students' individual work on sample exercises. In these settings, students generally work alone and in silence on exercises that require the rote application of memorized procedures, with teachers designating student answers as either correct or incorrect.

Not surprisingly, making new goals for students' mathematics learning a reality in classrooms has presented many challenges to teachers who have been asked to take on roles and responsibilities that are not consistent with their current teaching practices, their professional education, or their own experiences as students. In fact, the teachers themselves may have had limited opportunities to learn mathematics in a meaningful way. Although there is considerable consensus that meeting these challenges will require that teachers have deep insights about mathematics, about students as learners of mathematics, and about pedagogy that will support students' learning, there has been little consensus on how teachers should acquire this knowledge.

By design, this book appears at a time when the mathematics education community is renewing its commitment to reform. The introduction of *Principles and Standards for School Mathematics* (NCTM 2000a) represents a continuing effort to ensure that all students "learn important mathematical concepts and processes with understanding" (NCTM 2000a, p. ix). Building on the foundation of the original *Standards* document (NCTM 1989) and drawing on experiences in the field since its release, *Principles and Standards* provides a set of principles and standards that will fortify our efforts to create classroom environments that foster the development of mathematical power for all students.

It is now fitting that we begin to chart new directions for professional development, recognizing that teachers will not be able to meet the goals of reform without support that will help them deal with the challenges presented by teaching mathematics in new and unfamiliar ways. Toward that end, this book provides a new perspective on how to design, conduct, and evaluate professional education experiences for teachers that will support their efforts to improve their practice. The approach described herein situates the professional education of teachers "in practice." In this view, materials that depict the work of teaching (e.g., student work, mathematics instructional tasks, and classroom episodes) are used to create opportunities for critique, inquiry, and investigation. According to Ball and Cohen (1999, p. 12),

> Teachers can certainly learn subject matter, as well as knowledge of children, learning, and pedagogy, in a variety of courses and workshops. But the use of such knowledge to teach depends on knowledge that cannot be learned entirely either in advance or outside of practice.

By situating teacher learning "in practice," teachers have the opportunity to develop knowledge central to teaching by engaging in activities that are at the heart of a teacher's daily work. In this way, teachers develop knowledge through analysis of real situations.

This book, therefore, explores a specific type of professional development that connects the ongoing professional development of teachers to the actual work of teaching. This is not to say that all professional development in which teachers engage over the course of their careers should be practice-based. There are many other viable professional development strategies that may be appropriate at various points in a teacher's development for different purposes. For example, several of the strategies identified by Loucks-Horsley et al. (1998)—Immersion in Inquiry into Mathematics, Immersion in the World of Mathematicians, Curriculum Development, and Partnerships with Mathematicians in Business, Industry and Universities—can provide rich opportunities for teacher growth and development.

It is intended that the suggestions presented in this book will give professional development providers—mathematics teacher educators, teacher-leaders, staff developers, and supervisors—some practical assistance in their efforts to design, conduct, and evaluate professional development experiences that will have the potential to transform teachers' knowledge, beliefs, and habits of practice by making direct connections with the practice of teaching. If mathematics education reform is going to take hold on a large scale, we need to facilitate the process by creating opportunities that will "enrich teachers' capacity for understanding and intelligent decision-making" (Darling-Hammond 1993). The suggestions in this book provide a step toward that goal.

CHAPTER 1

MAKING THE CASE FOR REFORMING PROFESSIONAL DEVELOPMENT

The professional development of teachers is a key ingredient in improving our nation's schools (Sykes and Darling-Hammond 1999). The perceived importance of professional development is directly related to the ambitious nature of the reform goals and standards that have been put into place over the past decade by the National Council of Teachers of Mathematics (1989, 1991, 1995, 2000a), state departments of education, and the National Board for Professional Teaching Standards (1997). These documents call for students who can reason about challenging and complex problems that give rise to significant mathematical understandings. It also calls for teachers who can appropriately support students' learning by creating environments that foster communication, inquiry, and investigation. Meeting these goals and standards will require a great deal of learning on the part of teachers, the vast majority of whom were taught and learned to teach under a paradigm of instruction and learning in which memorization, repetition, speed, and correct answers were of paramount importance.

The kind of learning that will be required of teachers has been described as *transformative* (involving sweeping changes in deeply held beliefs, knowledge, and habits of practice) as opposed to *additive* (involving the addition of new skills to an existing repertoire) (Thompson and Zeuli 1999). Teachers of mathematics cannot successfully develop their students' reasoning and communication skills in ways called for by the new reforms simply by using manipulatives in their classrooms, by putting four students together at a table, or by asking a few additional open-ended questions. Rather, they must thoroughly overhaul their thinking about what it means to know and understand mathematics, the kinds of tasks in which their students should be engaged, and, finally, their own role in the classroom.

Teachers cannot be expected to undergo changes as profound as this—totally refurbishing their knowledge, beliefs, and habits of practice—on the basis of professional development as we know it. For most teachers in the United States, professional development includes mandated district-sponsored staff development and elective participation in courses, workshops, and summer institutes, often given by university-based teacher educators. District-sponsored staff development typically offers a menu of training options (workshops, special courses, or in-service days) designed to transmit a specific set of ideas, techniques, or materials to teachers (Little 1993). For example, teachers may be asked to select workshops from a list that includes training on the use of computers, cooperative group instruction, or assessment by portfolio. Such approaches treat teaching as routine and technical (Little 1993), encourage tinkering around the edges of practices rather than undertaking a total overhaul of practice (Huberman 1993), and may or may not relate specifically to the teaching of mathematics. In addition, they provide teachers with limited access to intellectual resources outside the teaching community and provide limited opportunities for meaningful collegial interactions within the teaching community (Little 1993).

Courses given by members of a university faculty are often associated with degree or certification programs and generally have an academic rather than an applied focus. These courses are often taught in a manner that is inconsistent with the ways in which we are asking teachers to teach or ways that will allow even the successful student to construct adequate or appropriate knowledge (Ball 1991; Silver 1994). Although workshops and summer institutes sponsored by teacher educators can be more practice-oriented, they generally include limited follow-up support for implementation. Like district staff development, one-day sessions or even two-week institutes usually do not take into account the positive and negative factors within the school environments to which teachers return and hence may have little impact on practice. In addition, both district staff development and university-sponsored workshops and courses are usually planned without any input from those for whom the professional development is intended (Fullan 1991). Generally, both of these types of professional development activities result in a disconnected and decontextualized set of experiences from which teachers derive additive benefits (i.e., the addition of new skills to their existing repertoires). They have not been designed to produce the kind of in-depth reexamination of beliefs that is necessary to inspire the changes required for the newer, more complex forms of teaching that are being recommended.

To support instructional change in mathematics, new forms of professional development are needed for teachers at all stages of their careers—forms that can affect teachers' actions and interactions in the classroom and lead to improved learning outcomes for all students. Little (1993) has argued that these new approaches should build teachers' capacity for complex, nuanced judgments about the process of mathematics teaching and learning. Others have argued that teachers must have, at a minimum, deep and flexible understandings of the mathematics that they will teach (Simon and Blume 1994; Thompson and Thompson 1996; Sowder et al. 1998). This is particularly critical since, as Ball contends, in the "reform" view of mathematics teaching, "the conception of content is more uncertain than a traditional view of mathematics as skills and rules, the view of children as thinkers more unpredictable" (1993, p. 394).

Several approaches to teacher professional development have been successful in assisting teachers in adopting reform-oriented instructional practice. For example, some have facilitated change by assisting teachers to learn new mathematical topics in ways that reflect the style of teaching and learning that reformers advocate for classrooms (e.g., Simon and Schifter 1991; Wilcox et al. 1991). Others have focused on helping teachers understand important nuances of students' mathematical thinking (e.g., Carpenter et al. 1989) and the ways in which intellectually productive social interactions can be developed and maintained in order to facilitate students' learning of key mathematical ideas (e.g., Cobb et al. 1991). What we have learned from the efforts of these and other researchers lays the foundation for the approach proposed in this book.

6

CHAPTER 2

SITUATING PROFESSIONAL
DEVELOPMENT IN PRACTICE

The first chapter argued that professional development as we now know it will not transform teachers' knowledge, beliefs, and habits of practice. In this chapter, we turn our attention to describing a program of professional development that has the potential to build teachers' capacity for innovative practice and ultimately to impact student learning. As indicated in the introduction, this book takes the stand that the professional development of teachers should be situated in practice. In this view the everyday work of teaching would become the object of ongoing investigation and thoughtful inquiry (Ball and Cohen 1999). Teachers would develop an understanding of subject matter, of pedagogy, and of students as learners—critical components of a teacher's knowledge base for teaching (NCTM 1991; National Board for Professional Teaching Standards 1997; Shulman 1986)—by investigating tasks that are central to teaching. Rather than learning theories and applying them to the practice of teaching, theories or general principles emerge from closely examining practice.

Hence, "samples of authentic practice"—materials taken from real classrooms—would become the curriculum for teacher education by providing opportunities for critique, inquiry, and investigation. For instance, a mathematical task along with a carefully selected set of student responses is one such example (Stylianou and Smith 2000). Teachers could be asked to complete the task, share various approaches that could be used to solve the task, and identify the mathematical ideas that are central to the task. The examination and analysis of student responses to the task could center on determining what students' responses reveal about students' mathematical understandings and misconceptions, the type of feedback that could be provided to specific students, and the questions that teachers could ask a particular student in order to better understand his or her way of thinking. Such a discussion is likely to enhance teachers' knowledge of mathematics content and of students as learners of mathematics.

These practice-based materials, however, are not self-enacting (Ball and Cohen 1999). Rather, they provide the raw material around which "professional learning tasks" (PLTs) can be designed (Ball and Cohen 1999, p. 27). PLTs, tasks that engage teachers in the work of teaching, can be developed in order to meet a specific goal for teacher learning and to take into consideration the prior knowledge and experience that teachers bring to the activity.

The Work of Teaching

The central tenet of this approach is that it is "centered in the critical activities of the profession—that is, in and about the practices of teaching and learning" (Ball and Cohen 1999, p. 13). One way to design professional learning tasks is to consider the cycle of teachers' work and the nature of the activities in which teachers engage as they move through the cycle.

The cycle begins with planning for instruction. Here the teacher decides what mathematical knowledge and processes she wants students to learn; determines the relevant prior knowledge and experiences on which students can draw to construct new knowledge; and creates, finds, or adapts tasks or activities that build on prior knowledge and experiences and have the potential to foster the intended learning.

The cycle continues with teaching—enacting the plan that has been developed. It is during the act of teaching that the teacher must engage students' in the task or activity, make midcourse corrections as needed to fit the needs of the students, provide "scaffolding" for students' learning so as to sustain their engagement in worthwhile mathematical activity, and formally and informally assess what students are learning.

The teacher completes the cycle with reflection. During this process, teachers must consider the level and kind of thinking in which the majority of students engaged during the lesson and what students did and said that suggested understanding of important mathematical ideas. They must also consider the ways in which the teaching may have supported or inhibited students' engagement with the task as intended. Based on an appraisal of the lesson and knowledge of the overarching mathematical goals, the cycle begins again with planning the next lesson.

Although this description may oversimplify the components of the teaching cycle, it serves to highlight the types of activities that are foundational to teachers' work and suggests potential PLTs that use authentic practice. For example, a videotape of a classroom episode could serve as the basis for several tasks that embody the work of teaching. Teachers could begin by analyzing the task that was used during instruction and by asking questions such as the following:

- What opportunities to learn mathematics are afforded by the task?

- What prior knowledge and experience would students need in order to engage in the task successfully?

- How would you expect students to go about solving the task?

Teachers could then move on to watching the video and analyzing the learning environment, responding to questions such as these:

- What decisions did the teacher make during the course of the lesson?

- What decisions were made by students?

- Who validated answers?

- Who asked the questions?

- What was the nature of the questions asked by students? By the teacher?

The investigation could continue with teachers analyzing what students seemed to be learning and how they learned it. Questions such as these might frame the analysis:

- What were the mathematical ideas with which students appear to grapple?

- What do students' solutions tell about what they know and understand?

- What factors appeared to support students' engagement in mathematical activity?

- What factors seem to hinder such engagement?

The discussion could conclude with actually planning the subsequent lesson, focusing on questions that include the following:

- What should be the mathematical target of instruction in the next lesson?

• What knowledge have students demonstrated that will serve as a foundation for constructing new knowledge?

• What task would accomplish the learning goal?

A videotape of teaching, therefore, could serve as the basis for engaging teachers in an investigation and analysis of all phases of the teaching cycle.

The videotape and students' work discussed so far represent two specific samples of practice-based materials that can serve as the basis for PLTs for teachers. The remainder of this chapter will focus on three broad categories of such materials—mathematical tasks, episodes of teaching, and illuminations of students' thinking. These materials will provide the foundation for PLTs that involve exploration and analysis.

Mathematical Tasks

Tasks used in the classroom form the basis for students' learning (Doyle 1988). Tasks that ask students to practice paper-and-pencil computations are likely to offer students one type of opportunity for thinking; tasks that require students to think about a situation rather than to follow a prescribed procedure offer them very different opportunities. According to Hiebert and his colleagues, "tasks that encourage reflection and communication are tasks that link up with students' thinking" (Hiebert et al. 1997, p. 20).

In planning for instruction, teachers must determine the mathematical concepts and processes they want their students to learn and then select tasks and activities that have the potential to promote the intended learning. Thoughtful planning requires (at a minimum) that teachers understand what mathematics children need to know in order to solve a task, recognize the mathematics embedded in a task, and match their goals for the students' learning with tasks that have the potential for achieving the goals.

The exploration of mathematical tasks provides teachers opportunities both to consider issues of the students' learning and to construct or reconstruct their own understanding of what mathematics is and how one does it. Consider, for example, the task shown in figure 1, adapted from the *Visual Mathematics Course* (Foreman and Bennett 1996). The task provides an opportunity for teachers to look for the underlying mathematical structure of a pattern, to use that structure to continue the pattern, and to develop a rule that can be used to describe and build larger figures. The task provides an interesting context for discussing what algebra is and how algebraic reasoning can be developed. (Teacher-generated solutions to this task can be found at www.cometproject.com.)

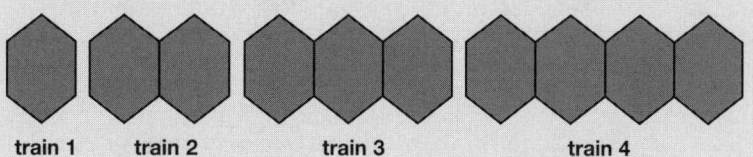

Hexagon Pattern Train

For the pattern shown below —

• compute the perimeter for the first four trains;

• determine the perimeter for the tenth train without constructing it; and

• write a description that could be used to compute the perimeter of any train in the pattern.

train 1   train 2   train 3   train 4

Fig. 1. An example of a task that could be the basis for exploration and analysis. (From Linda Foreman and Albert Bennett, *Visual Mathematics Course*. Copyright © 1996. Adapted by permission.)

Teachers might be asked to engage in several different professional learning tasks that involve the hexagon-pattern task. For example, teachers might—

- solve the task, discuss different strategies and approaches, and relate different methods to one another and to the visual representation;

- determine what prior knowledge and experiences students need to complete the task and what they might learn from it; or

- discuss how the hexagon-pattern task might contribute to the development of algebraic reasoning and how teachers might build on or extend the task.

Tasks can be drawn from challenging mathematics curricula (see Appendix A) appropriate to the grade level of the teachers, from assessments, or from a plethora of available instructional materials. Tasks can be selected so as to highlight specific processes (i.e., problem solving, reasoning and proof, communication, connections, or representations) or organized to show the development of mathematical ideas in a particular content area (i.e., number and operation, algebra, geometry, data analysis, and probability).

Episodes of Teaching

According to *Professional Standards for Teaching Mathematics* (NCTM 1991), a primary factor in teachers' professional growth is the extent to which they "reflect on learning and teaching individually and with colleagues" (p. 168). It is through reflection that teachers can gain insights into how their actions and interactions in the classroom influence students' opportunities to learn mathematics.

In order to develop the ability to reflect on their own teaching, teachers need opportunities to analyze and critique episodes of teaching, whether captured on videotape or CD, observed directly, or portrayed in written accounts. Consider, for example, the case of the "Ratio of Girls to Boys" shown in figure 2 (Barnett, Goldenstein, and Jackson 1994a, p. 122). In this episode, the teacher introduces the concept of ratio by comparing the number of girls to the number of boys in the class. The students confuse fractions and ratios and struggle to make sense of the comparison. The case provides an opportunity for teachers to explore important mathematical ideas related to rational number understandings and to consider how the teacher's actions and interactions in the classroom influenced what students learned.

The Ratio of Girls to Boys

We often hold discussions in my seventh grade class, and sometimes I find myself totally unprepared for the questions my students ask. It's not that I feel I should be the one with the answers, but I do want to guide the discussion productively.

In a recent lesson introducing ratio concepts, I had written the fraction 1/2 on the board and reminded my students that it meant "1 divided by 2," and that it also meant "1 out of 2."

"It is a division problem and also a ratio because it shows a comparison of 2 numbers—1 and 2," I explained. "Let's compare the number of boys and girls in our class." The class determined that there were 17 girls and 15 boys present.

"What is the ratio of girls to boys?"

Several students called out, "Seventeen to 15." I wrote the fraction 17/15 on the board.

Carmen blurted out, "That can't be right. You said a fraction means the top divided by the bottom. That'll be more than 1 whole class."

Laura interjected, "And you said you could say 'out of'—like 17 out of 15. That doesn't sound right—17 girls out of 15 boys."

I realized that I wasn't very clear myself about how fractions and ratios were related or what part context plays in describing ratios. These were good questions and I wasn't sure how to handle them.

Fig. 2. An example of a teaching episode presented in written form. (Reprinted from *Fractions, Decimals, Ratios, and Percents* edited by Carne Barnett, Donna Goldenstein, and Babette Jackson. Copyright © 1994 by Far West Laboratory for Educational Research and Development. Published by Heinemann, a division of Reed Elsevier Inc., Portsmouth, N.H. Used by permission of the publisher.)

Teachers might be asked to engage in several different professional learning tasks that center on reading and analyzing the case. For example, they might—

- discuss the different ways ratios can be expressed (1 out of 2, 1 to 2, 1 for every 2, 1 divided by 2), how these ways differ conceptually, and the differences between ratios and fractions (see Barnett, Goldenstein, and Jackson 1994b for additional insight);

- determine the sources of students' confusion and discuss the ways in which the confusion could be addressed; or

- plan a follow-up lesson to the one portrayed in the case.

Such opportunities invite teachers to make connections between the events depicted in the episodes and their own knowledge of mathematics and classroom practice. Unlike other materials, cases integrate elements that are often treated separately: content and processes, thought and feeling, and teaching and learning (Shulman 1992). (Smith [in press] also gives an extended example of the use of a case to promote teacher learning.)

Determining what students know and can do mathematically is a critical aspect of teaching, since it allows teachers to make appropriate instructional decisions before, during, and after teaching. To understand how students think about mathematics, teachers need to begin to understand mathematics from a child's point of view. Oral or written explanations, solutions, and reflections produced by students during instruction, in interview situations, or as on-demand performances in assessments provide insights into students' thinking.

Illuminations of Students' Thinking

Examining students' work affords the opportunity for teachers to move from their own understanding of a task to a broader consideration of what students' responses might reveal about their thinking, what difficulties such tasks might present for students, and how teachers might help students address (or avoid) common misunderstandings. According to Evans, "samples of student work are concrete demonstrations of what is known and what is not known" (1993, p. 72). Consider for example, the work of five students on the tree-tower problem shown in figure 3 (on next page) (Lamon 1999, pp. 1–2). The students' responses reveal varying levels of sophistication in their ability to reason proportionally.

Teachers might be asked to engage in several different professional learning tasks that center on analyzing the work produced by students. For example, they might—

- rank the responses according to the sophistication of their mathematical reasoning and explain the rankings (Lamon 1999);

- indicate what each response reveals about the student's ability to reason proportionally; or

- write a question that each student could be asked that would elicit the student's thinking, and explain what the question is intended to reveal about the students' conceptions.

13

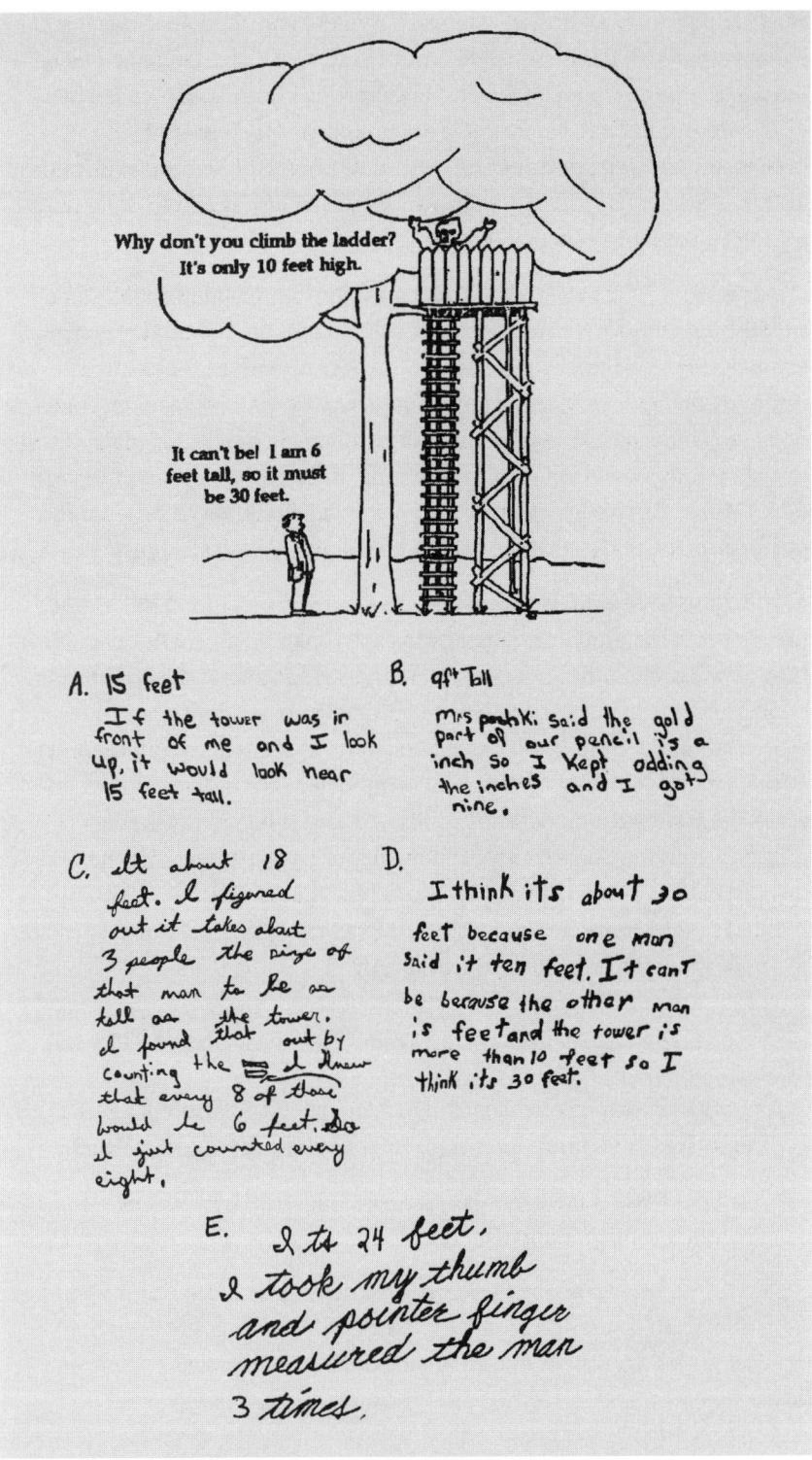

Fig. 3. Sample student responses to the tower task. (From Susan J. Lamon, *Teaching Fractions and Ratios for Understanding*. Copyright© 1999 by Lawrence Erlbaum Associates. Reprinted by permission.)

Examining students' work can help teachers realize that children's ways of interpreting, representing, and solving problems are different from the teacher's, but their methods may be equally valid. In addition, it can help teachers develop the ability to interpret or make sense of students' solution strategies and forms of representations. In a recent study, for example, students' work served as vehicle for facilitating teachers' learning about strategies that were unfamiliar to them and served as a springboard for discussions of ways to design instruction that would build on their students' thinking (Kazemi and Franke 2000).

Interviews with students can reveal students' thinking in more detail than can be obtained from an analysis of written responses alone. According to Ginsburg and his colleagues, interviewing "can provide insight into the distinctive ways in which children think about—construct—the world of school mathematics" (Ginsburg, Jacobs, and Lopez 1998, p. 16). Consider, for example, the interview excerpt shown in figure 4, in which Manny rolls two ordinary dice and is asked to give the sum of the two numbers rolled (Ginsburg, Jacobs, and Lopez 1998, p. 59). The task provides an opportunity for the teacher to gain insight into the strategies that her students use to solve addition problems.

Roll two dice and find their sum (Manny)

M: [Rolls 5 and 1.]

T: What numbers did you get?

M: 5 and 1.

T: How much did you get all together?

M: 6. [Rolls again, gets 3 and 3.]

T: What did you get?

M: 6.

T: How did you get that so quickly?

M: I know that 3 and 3 is 6.

M: [Rolls 6 and 3.]

T: Now?

M: 9.

T: What did you say to yourself in your head?

M: 6, 7, 8, 9.

T: Again.

M: [Rolls again.] 11.

M: [Rolls again.] 10.

T: How did you know that?

M: Last time I got 6 and 5. This time I got 6 and 4, so it has to be 1 less.

Fig. 4. Example of an interview conducted by a teacher. (From Herbert P. Ginsburg et al., *The Teacher's Guide to Flexible Interviewing in the Classroom.* Copyright © 1998 by Allyn & Bacon. Reprinted/adapted by permission.)

Teachers might be asked to engage in several different professional learning tasks that center on analyzing what Manny knows about addition. For example, they could—

- determine what Manny's responses to the dice task reveal about his understanding of addition and identify the different strategies he uses to solve the problems;

- analyze the questions asked by the teacher, identify those that seem to be generic and those that seem to be specific to the situation, and discuss which questions seem to elicit the most useful information; or

- discuss how the information gained from the interview with Manny could be used to plan instruction.

Interpreting interviews helps teachers to determine what line of reasoning and conceptual understanding is behind the procedures that students use to solve tasks, to plan for instruction, and to explore mathematics.

Summary

A practice-based approach to professional development provides teachers with an opportunity to develop new levels of awareness and knowledge through consideration of samples of authentic practice. The goal of such work is to help teachers develop the capacity to see specific events that occur in the practice of teaching as instances of a larger class of phenomena. That is, generalities are abstracted from examining particular situations, and these in turn become practical wisdom that will inform teachers' practice. Instead of learning theories and applying them later to practice, teachers witness the emergence of theories from the study of practice. Ultimately, the goal is for teachers to be able to apply these generalizations to their own practice.

Samples of authentic practice are not, however, a panacea for all the shortcomings of professional education. We must be aware of the potential pitfalls as well as the promise of these materials. Ball (2001) cautions us to consider carefully the way in which we design and conduct learning experiences for teachers around records of practice. In particular, she has identified four issues that should be taken seriously as we begin to construct learning experiences that build on accounts of practice:

1. *Curricula need to be developed around records of practice.* Isolated encounters with interesting professional learning tasks that do not build on each other to lead teachers to a more robust understanding will not facilitate fundamental changes in their practices. The same care that is given to developing curricula for children needs to be given to developing curricula that is for teachers and has records of practice at its core.

2. *The records of practice do not represent a teacher's own teaching situation.* Although this aspect of a practice-based curriculum can be a strength, the curriculum should be designed so that the professional learning tasks that center on records of practice are both relevant and compelling for teachers.

3. *Records of practice provide rich detail about particular situations, but it is important to see beyond details instead of becoming caught up in them.* Teachers need to be able to see specific events as examples of more generalizable ideas about mathematics teaching and learning. Moving from particulars to generalizations may not happen in one discussion. A sustained effort needs to be made to help teachers make these connections over time.

4. *The process of examining records of practice can become so analytical that it loses its connection to the work of teaching.* The analysis of records of practice is intended to help teachers develop a knowledge base for teaching that will improve their decision making in the classroom. The process is *not* designed to help teachers become more skillful at performing analysis for its own sake. Avoiding this pitfall requires keeping the work of teaching as a focus and making connections between the task at hand and the real work that teachers do.

The next chapter describes examples of professional development sessions that are based in practice.

18

CHAPTER 3

SNAPSHOTS OF PRACTICE-BASED
PROFESSIONAL DEVELOPMENT

This chapter provides concrete examples or snapshots of the approach described in chapter 2 in order to flesh out a vision of professional development that differs in kind from the more traditional forms discussed in chapter 1. The examples presented here exhaust neither the situations in which one might conduct or engage in professional development nor the samples of authentic practice that one could use to engage teachers in the work of teaching. From a practical perspective, the artifacts on which the examples are based represent what is currently commercially available or what might be easily gathered. It is likely that in time additional materials will become available, and thus the set of possibilities will expand. (See Appendix B for a list of materials that could be used as the basis for professional learning tasks (PLTs) similar to those described in this book.)

It is also important to note that each snapshot focuses on a specific professional learning task designed for a particular group of teachers with a specific goal in mind. Many other professional learning tasks that might be appropriate depending on the context, teachers, and goals for teacher learning could be designed using the same materials. The snapshots are in three categories: mathematical tasks, episodes of teaching, and illuminations of students' thinking.

On the first day of her elementary mathematics methods class, Elizabeth Travis presented her preservice teachers with the "Pirating Pizza" task, drawn from a sixth-grade unit in the *Connected Mathematics* curriculum (Connected Mathematics Project and Michigan State University 1996) and available through the Showme Center, an NSF-funded effort that focuses on middle school curriculum. The goal of the task is to determine how much pizza is left after a week if the pizza pirate eats one-half of the pizza the first night, half of what is left of the pizza on the second night, and half of the remainder on each successive night. Extensions of the task require students to determine the amount of pizza eaten on any night and whether the pirate will ever eat all the pizza. The task is mathematically rich, providing opportunities to explore the power of patterns in forming generalizations, to focus on the meaning of the numerator and denominator of fractions, and to give meaning to the multiplication of fractions.

The preservice teachers' encounter with the task was intended to call into question their assumptions about what mathematics is and how one does it. This, Elizabeth reasoned, was likely to occur because the selected task differed from the types of tasks that her students had previously experienced. It could not be solved by application of a single algorithm; the solution path was not immediately evident and involved exploration and reasoning through alternatives; multiple strategies could be used to solve the problem; and pictures or models could be made to clarify the situation. Students were encouraged to work together in small groups, to share their thinking about the problem with their peers, and to use diagrams, calculators, and other tools to support their work. The instructor did not validate solutions but rather asked for explanations and justifications as the preservice teachers worked through the problem.

Once all the groups had come up with an answer to the initial question, "How much pizza was left after a week?" Elizabeth asked groups to share their solutions and strategies with the class. Students were surprised to see that not everyone had solved the problem the same way. One group made a table and systematically recorded what happened to the pizza every night—how much was eaten each night, how much had been eaten in total, and how much was left. Another group folded a large sheet of newsprint repeatedly, labeling the fractional parts on each fold until they had accounted for seven nights. A third group "just knew" to multiply the amount left each night by 1/2 to find the amount eaten the subsequent night (e.g., night 1, $1/2 \times 1 = 1/2$; night 2, $1/2 \times 1/2 = 1/4$; night 3, $1/2 \times 1/4 = 1/8$). Students were then asked to identify patterns and to see if they could use these patterns to predict the amount of pizza that the bandit would eat on the tenth day, the hundredth day, and finally, on any day. After working on the extension in groups, students began to notice that the amount of pizza eaten on any day could be found by finding the value of the unit fraction in which the denominator was the product of the same number of twos as the number of days. This was eventually expressed as $1/2^n$, where n was defined as the number of days.

The second phase of the students' work on the "Pirating Pizza" task involved reflecting on their experience. Specifically, students were asked to reflect on the similarities and differences between their experience with the pizza problem and other mathematics experiences they had had over the years. The preservice teachers indicated that although the content (fraction multiplication) and the context (pizza) were similar to problems they had done in the past, nothing else was the same. For example, they indicated that the instructor wasn't "in charge" and didn't "tell us how to do it." The pizza problem could be done in many ways, not just "the teacher's way." Although they all agreed that the experience had its frustrating and anxiety-provoking moments, they also agreed that "it felt really good" when they were able to figure out and explain the pizza pirate phenomenon. Many commented that they had never worked in a group before and that it was really helpful to hear people talk about how they solved the problem and to be able to talk in a smaller forum than the usual whole-class configuration. They also appreciated having the time to figure out the problem. Speed was not what was valued.

Snapshot: The Human Chain Wrist Experiment

Charlie Ingram had been asked to work with a group of Hamlet School District high school teachers who were trying to decide how to implement the new state standards for mathematics that had recently been adopted. Charlie saw this as an opportunity to accomplish at least two goals. First, he wanted to give teachers an experience that would help them think about mathematics teaching and learning in a new way. He reasoned that if he could engage them in a challenging problem that embodied the mathematics they needed to teach, they might begin to rethink existing practices. Second, he wanted teachers to see how using mathematically rich tasks that focused on big mathematical ideas could serve as a springboard for exploration and discussion that would enable them to meet more than one standard.

Charlie decided to kick off his work with Hamlet teachers by having them engage in the "Human Chain Wrist Experiment," a task taken from *Cognitive Tutor Algebra 1* (Carnegie Learning 1999). This problem, calls on students to make a human chain with each student holding the wrist of the person to his or her right. When the teacher says, "Go," the first student gently squeezes the wrist of the next person, and so on through the chain. The last student says, "Stop," when he or she feels the squeeze on his or her wrist. Three designated timers record the time elapsed from when the teacher says go until the last student says stop. The average of the three times is then computed. The procedure is repeated with different numbers of students (i.e., chains of varying lengths), and the data are used to construct a graph of length of chain versus time. Students then use graphing calculators to find the linear regression line.

Charlie liked this problem for several reasons. It provided a concrete context in which to discuss ideas that often remain abstract and to focus on what linearity really means. Discussions of slope and intercept can be related directly to the experiment. The activity also provides a way of generating reliable linear data. No matter how many times Charlie has used the problem, the value for the regression coefficient has always been very close to 1. In addition, solving the problem can be facilitated by the use of a graphing calculator. Charlie has found that teachers who have not explored this technology find it a very useful tool in analyzing and making sense of the data. In his view, the calculator does the tedious work, thus "freeing up teachers" to focus on interpreting the values provided. The problem can also be extended by asking how long it would take to pass the squeeze through chains of varying lengths (e.g., 100 people, 100 000 people), how many people the squeeze could pass through in a specified amount of time (e.g., one hour, one day, one year); or by having the human chain pass the squeeze from shoulder to shoulder rather than from wrist to wrist and comparing the outcomes of the two experiments. Finally, Charlie has found that the problem is one that engages both teachers and students and therefore is one that teachers can actually take away and try in their own classes. He has found that if teachers have a successful experience in trying something new in their classrooms, they are more likely to be open to other new ideas.

The first day of their meeting, Charlie had the Hamlet teachers perform the experiment using ten different human chains. He then divided teachers into groups of four, trying to make sure that at least one group member had some experience using the graphing calculator. As they worked through the task, teachers discussed issues such as whether or not the x- and y-axes of the graph had to have the same scale, the units of measure for each variable in the linear regression equation, and the meaning of linearity. Once teachers had completed their analysis of the data, Charlie brought the whole group together to discuss the problem. There was some consensus among the teachers that the problem had been engaging and that it had made them think about the connections among the symbolic, graphic, and physical representations of a linear situation that they had not always taken time to consider.

Following the discussion, Charlie distributed copies of the new state mathematics standards and asked the teachers to decide which standards were addressed in the Human Wrist Experiment and to make specific connections to the problem. Teachers immediately identified the algebra standard that involved writing an equation for a line of best fit for a given set of data points and a related statistics-and-data-analysis standard that involved finding a regression equation of best fit. These tasks were direct matches with the question being asked in the problem. After additional discussion, however, the teachers also identified standards related to problem solving and communication, using the graphing calculator, and presenting the results of an experiment using a visual representation. Through the discussion, teachers began to see that a single problem, if rich enough, could accomplish multiple goals for student learning.

Considering the Payoff

By engaging in these types of mathematical activities, teachers often develop new understandings about a particular mathematical idea, make connections that they had not previous considered, and gain confidence in their ability to do mathematics. In addition, teachers may begin to develop an appreciation of mathematical tasks that can be solved in multiple ways and allow entry at various levels. They may also come to value their learning experience and want to give their students opportunities to learn mathematics in a way that is less centered on rules and procedures. Consider, for example, Karen Miller, a veteran middle school mathematics and science teacher who participated in QUASAR, a national project aimed at improving mathematics education for students attending middle schools in economically disadvantaged communities (Silver and Stein 1996). Karen attended SummerMath for Teachers (Schifter and Fosnot 1993) during the early stages of reconceptualizing her teaching practice. The opportunity to explore mathematical tasks during her SummerMath experience gave Karen a new perspective on teaching and learning. As Karen explains:

> It was a very hands-on situation where I made discoveries, I made mistakes, but even in my mistakes, I kept trying to put myself in the place of a student. In other words, the teacher became a student.... And I think that's extremely important.... I remember one time I pondered on an area and perimeter problem, and I kept playing with it and playing with it and playing with it and all of a sudden, I started laughing and [the instructor] came up and said, "What's up?" and I said, "I just discovered the formulas for area and perimeter.... How come it took me a whole day?" But the thing was, when I did get it, it was fantastic. I then thought, wouldn't it be wonderful if that's how children learned?... [then it] struck me that this is how I want **my** children to learn. I don't want to just sit there and give them algorithms.

The exploration of "good" tasks provides opportunities for learning and can be done in a relatively short time or can extend over a longer period, as in SummerMath for Teachers (Schifter and Fosnot 1993).

Connecting to Teachers' Own Practice

One approach to exploring mathematical tasks that focuses on the development of mathematical ideas in a content area involves using an innovative mathematics curriculum intended for children as the basis of professional development of teachers'. Acquarelli and Mumme (1996) and Ball (1996) note that curriculum materials can serve as resources for teachers' learning, especially when they are used in a process of inquiry and self-examination that is directed less toward adopting and implementing curriculum than toward using the curriculum as a tool for investigating problems of practice or examining new mathematical ideas or new representations of old ideas. There are some examples of professional development in which the core experience involves bringing in-service teachers in contact with innovative curriculum units (e.g., Acquarelli and Mumme 1996; Loucks-Horsley et al. 1998) and a few studies that show the impact of these types of experiences on teachers (Smith 1999) and on students' learning (Cohen and Hill 1998).

One such example is the case of Riverside Middle School (Smith 1997), one of the six urban middle schools that participated in the QUASAR project. Prior to implementing the *Visual Mathematics* (VM) curriculum (Foreman and Bennett 1996) in their classrooms, teachers participated in two courses that were specifically designed to introduce teachers to the VM materials. The overarching goal of the courses was to examine the role of visual thinking in understanding mathematical concepts and processes. This was accomplished primarily by providing teachers with opportunities to explore tasks themselves using the concrete materials and visual models, to explain orally how they solved problems that were posed, and to reflect on their experiences with the tasks. In addition, throughout the courses, teachers were routinely asked to think about how they would use the tasks with their students, to consider how their engagement with such tasks affected their thinking about mathematics teaching and learning, and to identify relationships or concepts that they understood better as a result of their experiences in the courses. In describing their experiences with the courses, one teacher commented, "Without those [courses] I would have had no basis on which to build. Those opened me up mathematically. I was very traditional."

The teachers' exploration of the curriculum provided them with a base of experience with specific tasks that they could and did use with their students. But they also derived something else from their engagement with cognitively demanding tasks and the rich discourse with colleagues and instructor—a general understanding of what high-level cognitive engagement is and how it might be encouraged.

Snapshot: Analyzing Cognitive Demands of Mathematical Tasks

Students in the Elmwood School District had not performed well on the constructed-response tasks that had appeared on the new state assessment. District leaders were concerned and decided to institute a series of monthly sessions for middle school and high school teachers in order to help them prepare their students better for the test. Wendy Carson, an experienced teacher educator in the area, was hired to work with the teachers. Wendy decided that looking at a diverse collection of mathematical tasks would provide a good starting point for their work together. She wanted to focus teachers' attention on different types of mathematical tasks and the different opportunities for learning associated with them. She wanted teachers to understand that for students to be successful in completing more challenging tasks on the state assessment, they needed to engage in such tasks throughout the school year.

When the twenty-five teachers arrived at the first session in late October, they were quickly assigned to groups and given a set of mathematical tasks. (Some of the tasks used in this activity are shown in figure 5.) The teachers were instructed to group together tasks that they considered the same in some way. The tasks varied with respect to their cognitive demands: some required repeating memorized facts (e.g., task H), some required using established procedures in routine ways (e.g., tasks D and F) or in ways that required a deeper understanding (e.g., tasks E and G), and others involved more nuanced thinking and reasoning (e.g., tasks A and C). The tasks also varied with respect to features that are often associated with reform-oriented instructional tasks. For example, some tasks required an explanation or description (e.g., tasks A, C, D, and G), some could be solved using manipulatives (e.g., tasks A, E, and F), some had real-world contexts (e.g., tasks B, C, and D), some involved multiple steps, actions, and judgments (e.g., tasks A, B, C, D, E, and G), and some made use of diagrams (e.g., tasks A, E, F, and G).

TASK A

Manipulatives/Tools: Counters

For homework Mark's teacher asked him to look at the pattern below and draw the figure that should come next.

Mark does not know how to find the next figure.

A. Draw the next figure for Mark.

B. Write a description for Mark telling him how you knew which figure comes next.

QUASAR Project - QUASAR Cognitive Assessment Instrument—Release Task

TASK B

Manipulatives/Tools: None

Part A: After the first two games of the season, the best player on the girls' basketball team had made 12 out of 20 free throws. The best player on the boys' basketball team had made 14 out of 25 free throws. Which player had made the greater percent of free throws?

Part B: The "better" player had to sit out the third game because of an injury. How many baskets (out of an additional 10 free-throw "tries") would the other player need to make in order to take the lead in terms of greatest percentage of free throws?

Adapted from *Investigating Mathematics*. Glencoe Macmillan/McGraw-Hill, New York, 1994.

TASK C

Manipulatives/Tools: Calculator

Your school's science club has decided to do a special project on nature photography. They decided to take a few more than 300 outdoor photos in a variety of natural settings and in all different types of weather. They want to choose some of the best photos and enter the state nature photography contest. The club was thinking of buying a 35mm camera, but one member suggested that it might be better to buy disposable cameras instead. The regular camera with autofocus and automatic light meter would cost about $40.00 and film would cost $3.98 for 24 exposures and $5.95 for 36 exposures. The disposable cameras could be purchased in packs of three for $20.00, with two of the three taking 24 pictures and the third one taking 27 pictures. Single disposables could be purchased for $8.95. The club officers have to decide which would be the better option and justify their decisions to the club advisor. Do you think that they should purchase the regular camera or the disposable cameras? Write a justification that clearly explains your reasoning.

TASK D

Manipulatives/Tools: None

The cost of a sweater at a department store was $45. At the store's "day and night" sale, it was marked 30% off of the original price. What was the price of the sweater during the sale? Explain the process you used to find the sale price.

TASK E

Manipulatives/Tools: Pattern Blocks

1/2 of 1/3 means one of two equal parts of one-third.

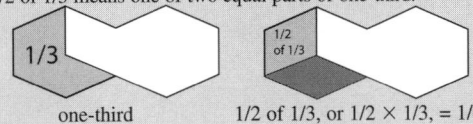

Find 1/3 of 1/4. Use pattern blocks. Draw your answer.

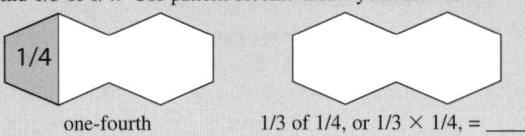

Find 1/4 of 1/3. Use pattern blocks. Draw your answer.

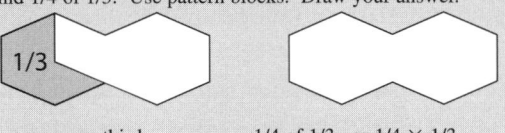

TASK F

Manipulatives/Tools: Square Pattern Tiles

Using the side of a square pattern tile as a measure, find the perimeter of (or distance around) each train in the pattern block figure shown.

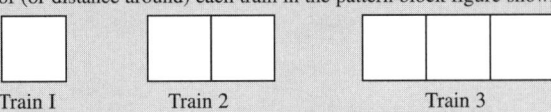

TASK G

Manipulatives/Tools: Grid Paper

The pairs of numbers in (a)–(d) represent the heights of stacks of cubes to be leveled off. On grid paper, sketch the front views of columns of cubes with these heights before and after they are leveled off. Write a statement under the sketches that explains how your method of leveling off is related to finding the average of the two numbers.

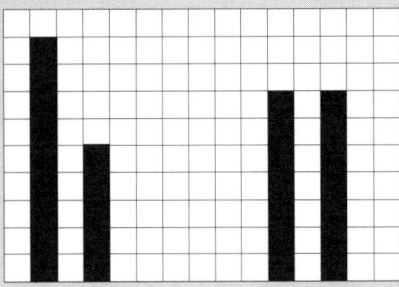

(a) 14 and 8   (b) 16 and 7   (c) 7 and 12   (d) 13 and 15

By taking two blocks off the first stack and giving them to the second stack, I've made the two stacks the same. So the total # of cubes is now distributed into 2 columns of equal height. And that is what average means.

Taken from Bennett and Foreman (1989/1991)

TASK H

Manipulatives/Tools: None

Give the fraction and percent for each decimal.

0.20 = _____ = _____.
0.25 = _____ = _____.
0.33 = _____ = _____.
0.50 = _____ = _____.
0.66 = _____ = _____.
0.75 = _____ = _____.

Fig. 5. Sample tasks from the sorting activity. (From Smith and Stein [1998].)

Although teachers' initial efforts to categorize tasks focused on surface-level features, producing such general types as visual tasks, fraction tasks, and word problems, further discussion made it clear that an analysis of the tasks had to go beyond superficial aspects to focus on the kind of thinking that students must use to complete the task. Teachers were then asked to categorize the tasks again, this time using one of four levels of cognitive effort: memorization, use of procedures without connections to concepts or meaning, use of procedures with connections to concepts or meaning, and mathematics (Smith and Stein 1998; Stein et al. 2000).

For example, Task F was originally put in categories designated by the groups as visual, computation and facts, or geometry. After some additional discussion within groups, teachers decided that the task should be classified as involving procedures without connections to meaning. They argued that "you really didn't have to think much because it told you what to do and how to do it." Another group argued that "you really didn't have to understand perimeter to do the task."

The sorting activity encouraged teachers to think deeply about what made a task challenging for students and to understand fully that one actually has to do a task to appreciate its complexity. They also decided that what might be challenging for one group of students might be routine for another, so teachers need to consider students' prior knowledge and experiences with tasks. By the end of the session, teachers were beginning to wonder how they would evaluate their own tasks. One teacher commented, "I think most of my tasks are low-level."

Considering the Payoff

Although selecting good tasks does not guarantee student engagement at a high level, it appears to be necessary to such engagement, since low-level tasks almost never result in high-level engagement. Hence, the long-term goal of this type of sorting activity is to raise teachers' awareness of how mathematical tasks differ with respect to their cognitive demands, thereby allowing teachers to match tasks more precisely to their goals for student learning and to become more analytic and reflective about the role of tasks in instruction. For students to be successful on assessments that require high-level thinking and reasoning, they need ongoing opportunities to engage in such tasks. Research has shown that students who participate in classrooms where good tasks consistently engage them in high-level cognitive activity make greater learning gains on tasks that require thinking, reasoning, and problem solving (Stein and Lane 1996).

Connecting to Teachers' Own Practice

Once teachers have some experience in identifying the cognitive levels of a predefined set of tasks, they may benefit from additional activities that more closely relate to their own practice. For example, in a follow-up to the session just described, Wendy Carson asked teachers to collect the tasks they used during classroom instruction over the following month. Her plan was to have teachers evaluate the tasks using the four categories and to discuss whether they provided sufficient opportunity for developing thinking, reasoning, and problem-solving skills as well as other, more basic skills. Another possible follow-up activity would be to have teachers evaluate the tasks in a unit or chapter of their textbook or other instructional materials. This could lead to a discussion of the balance of instructional tasks provided by the materials and to a consideration of how to rewrite low-level tasks so as to provide additional challenge to students.

Records of Teaching

Snapshot: Analyzing Fran's and Kevin's Teaching

By the twelfth week of their graduate-level advanced methods course, the practicing elementary and middle school teachers had participated in several personal mathematical experiences that had convinced them that concrete materials would help their students learn mathematics. Although manipulatives are certainly helpful in fostering students' construction of mathematical ideas, the instructor, William Jackson, wanted teachers to understand that manipulatives are not *inherently* good—it is the way in which manipulatives are used that matters most. (Stein and Bovalino 2001).

William decided to ask teachers to read "Multiplying Fractions with Pattern Blocks: The Case of Fran Gorman and Kevin Cooper" (Stein et al. 2000). He had three reasons for his choice: the mathematical content of the case—fraction multiplication—was the class's current focus, work to date had illuminated teachers' difficulties in explaining either in words or pictures what it means to multiply two fractions, and the situation presented in the case was likely to call into question some assumptions about the use of manipulatives. In the case, Fran and Kevin, who worked together in the same urban middle school, had planned a lesson on fraction multiplication using pattern blocks. Students were given exercises similar to those shown in figure 6 and asked to find the answers using pattern blocks and to draw their answers. Although both Fran and Kevin used the same task and started the lesson in much the same way, they supported student learning in very different ways. When Fran's students were confused, she asked them to start over and led them through a series of steps that resulted in the correct answer. In effect, Fran gave students a new procedure—one that involved pattern blocks—and the emphasis shifted from a focus on understanding to a focus on the correct answer. Kevin approached the situation somewhat differently. When his students were confused, he encouraged groups to work together to resolve their differences and then share their discussion with the class. He tried to ask questions that were based on how his students were thinking, not on his own understanding of the task. He continued to focus on connections between the representations and the meaning of fraction multiplication.

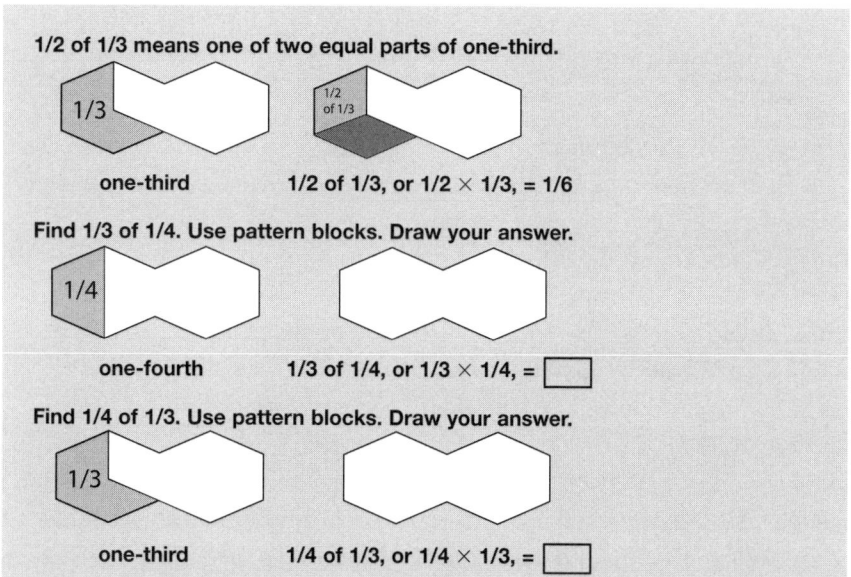

Fig. 6. Sample tasks used by Fran and Kevin. (From Matthew Zullie, *Fractions with Pattern Blocks*. Copyright © by Creative Publications. Reprinted by permission.)

Teachers' work with the case involved two components: first, they completed the task and discussed the mathematical ideas embedded in the task, and then they identified the similarities and differences between the two classes. Teachers were able to identify many important mathematical ideas that were embedded in the task that Fran and Kevin selected. These included the fact that a whole can be more than one thing, that the size of the fractional part depends on the size of the whole, that when multiplying two proper fractions the product is smaller than either factor, and that the order of multiplication doesn't matter (i.e., $1/2 \times 1/3 = 1/3 \times 1/2$). When discussing the similarities and differences between what Fran and Kevin had done, the teachers indicated that Fran had used manipulatives to teach traditionally—they were just another procedure to get the answer. Kevin, they argued, helped students develop an understanding of what it means to multiply fractions. Fran continued to go back to her way of thinking in working with students, whereas Kevin tried to build on how his students were thinking about the problem.

Snapshot: Analyzing Mrs. Joseph's Approach

The first-, second-, and third-grade teachers at the Fairmont Elementary School had been meeting biweekly during the semester with Mark Soloman, the district math supervisor, who was trying to support their efforts to implement a new curriculum. The teachers appreciated having the time to work with Mark and their colleagues. They often got practical suggestions on problems they had experienced as well as support and encouragement.

In their previous discussions, it had become clear that the teachers felt that they needed to show their students exactly how to do addition and subtraction so that they could solve these basic problems quickly and accurately. Mark wanted teachers to consider the different ways in which children might solve a problem without being told and decided to show a video in which children in a second-grade class were given the problem $26 - 17$ (Kamii 1989). In this video, Mrs. Joseph, the classroom teacher, asks several different students to share with the class how they solved the problem. In all instances, the students began by subtracting the tens ($20 - 10 = 10$). At this point, students' solution paths diverge—some students subtract 7 from 10 and add 6; others add 6 and then subtract 7. One student made an error and arrived at an answer of 11. Throughout the episode, Mrs. Joseph did not validate students' answers or suggest any methods for finding the solution. She left it up to the students to determine the validity of the approaches and solutions that were suggested. She had created a culture in her classroom where students were expected to ask questions and critique the work of their peers, and they did so without hesitation. Through this process of evaluating the work of their peers, students who did not initially arrive at a correct answer were able to revise their solutions.

When teachers were initially asked to consider how Mrs. Joseph supported students' learning, some seemed to think that she had played no significant role. At Mark's suggestion, they viewed the tape again, looking specifically at the problem Mrs. Joseph posed, the questions she asked, and the way she involved the students. This time, they understood the teacher's role better. Some teachers started to identify aspects of the classroom environment that had not been immediately obvious to them: the teacher had created a classroom where proof was important but justification and validation came from the students, not the teacher. She had posed a problem that provided an opportunity for students to think, and she had provided sufficient time for students to work on the problem. Moreover, she had encouraged invented procedures that focused on understanding. A few teachers believed strongly that Mrs. Joseph must have taught her students the various methods they used and questioned whether this was a reasonable use of instructional time. Others thought that the students' strategies were built on their current understandings and had been invented by the students, not learned from the teacher. Although the discussion ended without resolving this difference of opinion, the teachers began to wonder how their students would solve problems for which they did not have rules and procedures available.

Considering the Payoff

Examination of actual classroom cases is a particularly promising means of facilitating the development of content knowledge and supporting inquiry into classroom practices (Merseth and Lacey 1993). Barnett (1991) found that text-based cases had the potential for enhancing in-service mathematics teachers' pedagogical thinking and reasoning skills. Friel and Carboni (2000) found that cases presented on videotape facilitated preservice teachers' development of a more student-centered pedagogy.

Although episodes of teaching can generate much discussion about teaching, learning, and mathematics, the real value of examining these records of practice is in what teachers take from these experiences and apply to their own practice. In order to learn from examples and to transfer what has been learned in one situation to learning in similar situations, teachers must learn to recognize specific situations as instances of something larger and more generalizable. In order for teachers to be able to move from a specific instance of a phenomenon to a more general understanding, they need multiple opportunities to consider the issue, to make comparisons across situations, and ultimately to examine their own practice. For example, what teachers can begin to abstract from situations such as those of Fran and Kevin are ways in which they can support or inhibit students' learning of mathematics through their actions and interactions as students engage in challenging tasks.

Whether a teaching episode is presented in written or video format, it is important that any analysis of teaching focus teachers' attention on specific aspects of the situation that relate to goals set for their development (e.g., learning mathematics, considering students' thinking, analyzing teachers' decision making). This focus provides teachers with a lens through which to view events that unfold during a lesson—a criterion that signals what they need to pay attention to and what they can ignore. Written cases have the benefit of allowing teachers to focus on specific aspects of a lesson. The authors have made conscious decisions about what to give attention to and what to ignore. This decision eliminates some of the messiness that is endemic to classroom life in favor of a sharper focus on an issue or a set of issues that are the intended target of inquiry. In contrast, videotaped teaching episodes bring the classroom to life with all its complexities. They provide vivid images of what classrooms look like, feel like, and sound like. The question for teachers should not be "Which is better?" but rather "Which is the best match, given my goals for teacher learning at this specific point in time?"

Connecting to Teachers' Own Practice

Although commercially available cases and videos provide an important starting point for analysis of teaching, having teachers write their own cases based on episodes from their own teaching (Schifter 1996a, 1996b) or having them analyze videotapes of their own teaching are important ways to begin to connect ideas about teaching and learning with their own practices (Smith 2000a). These teacher-written or teacher-produced scenarios could be used for private reflection as well as for ongoing discussions among colleagues about teaching.

Illuminations of Students' Thinking

Snapshot: Analyzing Students' Responses to the Product Task

The "extend product pattern" task, which appeared on the National Assessment of Educational Progress in 1992, provided a good starting point for elementary teachers' consideration of issues related to students' thinking. (The task and thirteen student responses appear in Kenney and Silver [1997].) The preservice and practicing teachers who explored student thinking in the context of this task were enrolled in a master's course at a local university. The preservice teachers had completed their undergraduate degrees and were enrolled in a yearlong MAT program that would result in elementary certification. Some of the practicing teachers were working on master's degrees, and others were trying to accumulate the professional development hours mandated by the state.

In the "extend product pattern" task, fourth-grade students were given the pattern shown in figure 7 and asked if 375 could be one of the products. Students were asked to answer yes or no and to provide an explanation for why 375 would or would not appear in the list. Although 75 percent of students answered no, only about 25 percent wrote a mathematically correct rationale (Blume and Heckman 1997).

Extend Product Pattern

$$2 \times 2 = 4$$
$$2 \times 2 \times 2 = 8$$
$$2 \times 2 \times 2 \times 2 = 16$$
$$2 \times 2 \times 2 \times 2 \times 2 = 32$$

Fig. 7. The NAEP "extend product pattern" task

Before they discussed the students' responses, the teachers solved the problem, presented several different ways to approach it, and discussed what students needed to know and be able to do in order to answer the question. In their initial review of the students' responses, teachers were asked to consider what a student understood, what they were not sure that the student understood, and what questions they would like to ask the student in order to clarify the way in which he or she was thinking about the problem.

Many of the responses provided little information about what the fourth graders were actually thinking (e.g., " 'cause 375 cannot fit in," "because the pattern doesn't go like that," "because it would throw the pattern way off"). Even responses that provided some information on what the students understood (e.g., "2 is even and 375 isn't," "because 375 is not divisible by 2," "because 2 times as many 2s as you want does not add up to 375") raised many points that the teachers wanted to clarify. Teachers generated questions that they wanted to ask the students to help them learn more about the students' mathematical understandings.

Several important points were raised in the discussion. The teachers decided that they needed to be careful not to interpret what students said in light of their own understanding of the problem. For example, one student indicated that "it goes to 256 + 512" (Blume and Heckman 1997). One teacher argued that the student realized that continuing the sequence from the products given would result in 64, 128, 256, and 512. The student recognized, she argued, that 375 was between 256 and 512 and therefore would not be in the sequence. It was agreed that this could in fact be what the student meant, but that it might clarify the *student's* understanding if he or she were asked to explain his or her answer (e.g., "Where did the 256 and 512 come from? Why are you adding them together? What is the 'it' that 'goes to 256 + 512'?").

Another point that came up was the value of using multiple forms of assessment in order to determine what a child really knows. The teachers had read about Susie (Parker and Picard 1997), a student who did well in class and on verbal tasks but did not perform well on standardized tests. The two forms of assessment provided different portraits of what she knew. Teachers pointed out that tasks such as the NAEP product task, which asked for an explanation, provided some information about students' thinking that multiple-choice and yes-or-no options do not and that talking with and listening to a child provides yet more information. They argued that over time information from a variety of sources was helpful.

Teachers left the session with the assignment to interview two fourth-grade students about the task and to analyze what each child understood about the mathematics. The teachers planned to meet again to discuss what the interview added to their knowledge about the students' understanding.

Snapshot: Analyzing Responses to the "Extend Tiles Pattern" Task

The middle and high school mathematics teachers in River City were attending a summer in-service session that focused on algebraic thinking. The overarching goal was to help teachers begin to think about algebra not as a set of rules and procedures but as a way of understanding and generalizing relationships. The mathematics leadership in the district recognized the importance of making algebra accessible to all students and of helping teachers in both the middle and high school to see algebra as a way of thinking and reasoning, not just a set of procedures to be performed. They also wanted teachers to become more reflective about their teaching and the impact of their actions and interactions on students' opportunities to learn. Thus, over a five-day period teachers explored mathematical tasks, analyzed videotaped and written records, and reviewed students' work, all of which were intended to provide an alternative view of what it means to teach and learn algebra.

One component of the session involved exploring the pattern task shown in figure 8 and reviewing work produced by students. This task appeared on the National Assessment of Educational Progress in 1992 and 1996. See Silver, Alacaci, and Stylianou (2000) for a discussion of students' performance on this task and for several students' responses to this task. The first three figures in the pattern of tiles that contains fifty figures were given. Students were asked to describe the twentieth figure in this pattern, including the total number of tiles it contains and how they were arranged. They were also asked to explain the reasoning that they used to determine this information and write a description that would apply to any figure in the pattern.

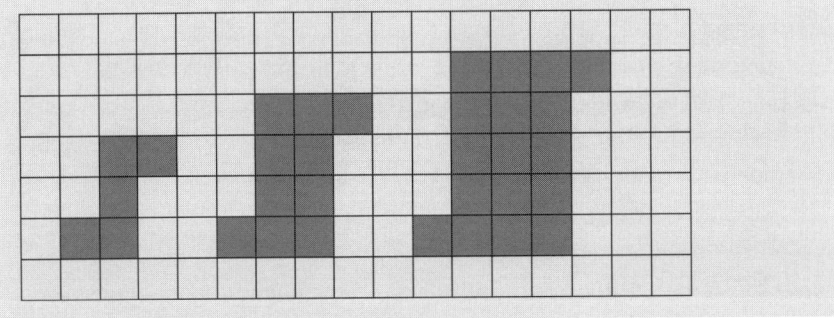

Fig. 8. The NAEP "extend tile pattern" task

The teachers began by exploring the task and identifying the mathematical ideas that were at the heart of the task. The high school teachers initially didn't see the task as having much to do with algebra. As Barbara Stapleton and Catherine Lawton—graduate students from a local university who had agreed to work with the River City teachers—began to ask specific questions about the task, the teachers began to see that *variable, generalization*, and *recursion* were important ideas embedded in it.

At this point, Barbara and Catherine suggested that teachers evaluate the set of responses using the five-level scoring rubric that the district used in evaluating the performance assessment that was given in the spring of each school year. The exploration of students' work raised questions regarding the need to express the general description as a formula and the giving of credit for correct solutions that did not have general descriptions or for general descriptions that did not include a formula. This served to generate further discussion about the criteria at each score level, how they applied to the specific task, and how teachers should weigh various responses.

The responses also provided an opportunity to discuss different approaches taken by students in solving the problem, the types of errors that students made in completing the task, and the conceptions that might underlie the errors. The responses, teachers ultimately decided, revealed a lot of evidence of algebraic thinking even though it was not always conveyed in a conventional way.

Considering the Payoff

Examining students' work can serve many purposes; overall, it can help teachers develop the capacity to make judgments about what students know and can do. Jackie, a student who was nearing completion of her certification program in secondary school mathematics, makes this point clear in her comment:

> *You really had to interrupt what they [the students] were saying, and then from there determine, not that they solved the problem, but if they understood what they were doing, and what you classify as understanding. Being able to read the students' work, trying to figure out how they are thinking—if you can't do that, you are in big trouble....*

Examining students' work can also expand teachers' views of what students can do when given the opportunity. For example, teachers may be surprised to see that students solved the NAEP pattern task in such interesting ways, and they may begin to wonder how their own students might respond. In addition, reviewing students' work moves teachers beyond their own ways of knowing and understanding mathematics to consider what students' approaches reveal about their understanding.

Connecting to Teachers' Own Practice

After teachers have had an opportunity to discuss and analyze students' responses, they could be encouraged to administer a task (perhaps one that has been discussed or a similar one) to their students as a preassessment or postassessment. They could then determine what the responses reveal about students' understanding or misunderstanding of the concepts and ideas that are embedded in the task and consider how they would design instruction that builds on students' current knowledge and understanding. This work could then provide a basis for ongoing discussions among teachers, such as those described by Moon (1997), whose framework may be helpful in creating study groups to examine students' work.

The examples in this section depict specific ways in which authentic practice can be used to create professional learning tasks for teachers in formal professional development settings. These examples highlight the materials used, the professional learning task in which teachers engaged, and a general sense of what the teachers gained from these experiences.

The types of formal learning opportunities described in this section represent a particular type of professional development in which members of a teacher community—defined by the school or district in which teachers work or by teachers' enrollment in a teacher-education or certification program—come together to engage in a pre-planned set of activities often involving someone from outside the community (e.g., university teacher educator, district staff developer). This is not to suggest that professional development cannot or does not occur in less formal or collaborative settings. For example, professional development happens in classrooms as teachers enact new instructional programs with their students and reflect on their experiences (Russell 1997). It happens when teachers leave formal professional development events and continue their conversations about teaching and learning with their colleagues and support providers in Web-based discussion groups or electronic mail exchanges. Recent efforts, such as the Internet Learning Forum at Indiana University, for example, have been designed to support and study the creation of online communities of teachers engaged in critical reflection on their instruction. In addition, professional development also occurs as teachers "tag up" (Stein, Silver, and Smith 1998) in informal exchanges that occur between and among teachers before and after school, during lunch breaks, and between classes.

Regardless of the setting in which professional development occurs, it is important to remember that it is the content of the professional development experience that matters most (Kennedy 1999). Teachers' learning is analogous to children's learning in that it is the nature of the activities in which they are asked to engage that will determine what they learn and how they learn it.

Practice-based materials have the potential to engage teachers in important tasks related directly to teaching and appear to have several advantages. They help ground discussions of abstract ideas, portray the complexities and dilemmas of teaching, and connect mathematics content, pedagogy, and students' thinking about mathematics. In addition, practice-based materials provide a common experience for teachers to refer to, discuss, and analyze that has the advantage of being grounded in teaching but not in *their* teaching. Finally, this approach is flexible. It can be used in both preservice and in-service settings or in settings that include both preservice and in-service teachers. It can be done in formal, organized settings as described in the snapshots or in less structured ways.

In the next two chapters, we turn our attention to a broader set of issues related to designing and assessing the professional development experiences. The discussion in these chapters, although relevant to practice-based professional development, is also applicable to other forms of professional development. These chapters highlight important considerations for those who are creating experiences for teachers, and they delineate ways of determining the effectiveness of those experiences.

38

4

DESIGNING HIGH-QUALITY
PROFESSIONAL DEVELOPMENT
EXPERIENCES

Now that practice-based professional development has been explained and specific examples have been provided, we turn our attention to broader issues. In particular, this chapter focuses on issues for consideration in the creation of transformative learning experiences for teachers.

The Professional Development Design Framework developed by Susan Loucks-Horsley, Peter Hewson, Nancy Love, and Katherine Stiles (1998) suggests a process by which good professional development programs in mathematics are created, implemented, and modified. Central to the design process is the view of teachers as both "agents and objects of their professional growth" (p. 260); the recognition that teachers need to be supported throughout the teaching cycle, which includes building knowledge, translating knowledge into practice, teaching, and reflecting on teaching; the consideration of the context within which the professional development will occur (e.g., who the students and teacher are, the current state of practice, the learning environment); and attention to the critical issues that influence the success of a professional development plan (e.g., leadership, professional culture). Hence, professional development is seen as a contextual, collaborative, and ongoing venture, involving both teachers and professional developers and intended to promote teacher growth and development over time. The framework, therefore, can provide a useful tool in the planning process, illuminating many important factors that may otherwise be hidden from view (Stein, Smith, and Silver 1999).

A central consideration in designing professional development experiences for teachers should be determining what one wants teachers to learn, both in short-term and long-term goals. This cannot be stated too strongly. Decisions should be based on an assessment of what teachers need or want to learn and what knowledge, skills, and experiences they bring to the enterprise. With this information in hand, professional developers can select appropriate materials and create professional learning tasks that build on teachers' prior knowledge and have the potential to foster the intended learning.

Another consideration is the strategy or setting for conducting formal professional development. Although there are many possibilities— workshops, courses, seminars, summer institutes, study groups, and department meetings—it is important to note that it is not the structural or organizational features of professional development experiences that will determine their effectiveness. Rather, it is the content of the programs that matters the most. A course, for example, can be a significant learning opportunity or a dreadful experience, depending on the content and the manner in which the course is conducted. According to Kennedy (1999, p. 6),

> A program whose content is not valuable will not be improved by increasing the number of contact hours, distributing contact hours over time, providing in-class visits, and so forth. Structural features alone provide no guarantee of improved teacher learning or of eventual benefit to students.

Other features are also important to consider in designing professional development experiences for teachers. Research over the last decade has identified the limitations of current professional development practices (e.g., Fullan 1991) and the salient characteristics of effective professional development (e.g., Sparks and Loucks-Horsley 1990; Darling-Hammond and McLaughlin 1995; Loucks-Horsley et al. 1998). In addition, research has provided insights on teacher learning (e.g., Schifter and Fosnot 1993; Wood, Cobb, and Yackel 1991) and has shown the impact of professional development on student learning (e.g., Carpenter et al. 1989; Cobb et al. 1991). From these and related studies, a set of features is beginning to emerge that can guide efforts to create new forms of professional development that will build teachers' capacity for innovative practice and improve student learning outcomes.

The remainder of this chapter discusses features that should be considered in designing professional development. When possible, the snapshots that were introduced in chapter 3 are used to illustrate the ways in which specific features are considered, implemented, and allowed to play out in professional development. In instances where the feature may not be apparent, given the brief nature of the snapshots, examples from other sources are provided.

Improving students' learning outcomes requires that professional development experiences be grounded in the belief that "excellence in mathematics education requires equity—high expectations and strong support for all students" (NCTM 2000a, p.12). Accordingly, professional development must provide opportunities for teachers to consider the impact of their teaching practices on what students learn and how they learn it and how *all* students—regardless of previous experience, race, culture, language, or gender—can be supported in learning sound and significant mathematics.

Professional Development Should Have Students' Learning as the Ultimate Goal

Professional development must therefore focus on ways to enhance students' learning. Too often professional development focuses narrowly on changing teaching behaviors (e.g., on helping teachers learn to use a new manipulative or piece of technology) with no attention to the impact of such materials on what students know and can do. Although teachers need to learn to use new techniques and tools, they also need to step back from their own learning and consider the implications for the students' learning.

William Jackson's use of the case of Fran and Kevin in his advanced methods class, as discussed in chapter 3, provides a useful example. Although Jackson wanted teachers to explore the use of pattern blocks in solving problems that involved the multiplication of fractions, this was not his only goal. He also wanted teachers to move beyond their own experience with the task and tools to consider how the use of the tool could support students' learning of mathematics. In addition, he wanted teachers to begin to consider the factors that might contribute to the successful use of manipulatives, since manipulatives in and of themselves do not promote students' learning (Stein and Bovalino 2001).

Professional Development Should Support the Ongoing Work of Teaching

Teachers need assistance that is connected to the actual practice of teaching. Such assistance must "engage teachers in concrete tasks of teaching, assessment, observation, and reflection that illuminate the processes of learning and development" (Darling-Hammond and McLaughlin 1995, p. 598). The approach described in chapters 2 and 3 of this book provides one way to support the ongoing work of teaching. In this approach, professional learning tasks in which teachers engage are deeply rooted in the practice of teaching. The teaching that is being examined, however, is not *their* teaching. Although this has many advantages, as discussed in the previous chapter, the ultimate goal of such work is to help teachers develop analytical tools and insights that can ultimately be used in consideration of their own practice.

Teachers also need assistance that focuses directly on their day-to-day efforts to teach in new and demanding ways. This type of assistance can be provided by supportive ventures that focus directly on an individual teacher's practice, such as coteaching, coaching, assistance with planning, and reflection on actual lessons (Schifter and Fosnot 1993). Another type of support for the ongoing work of teaching that has received considerable attention as a result of the Third International Mathematics and Science Study (TIMSS) video study is lesson study. In this approach, groups of teachers meet regularly over long periods of time to design, implement, test, and improve lessons (Stigler and Hiebert 1999). Both forms of assistance aim to stimulate teachers to reflect on their current practice so as to enhance teacher-planning and decision-making processes.

Professional Development Should Be Grounded in Mathematics Content

Teachers must understand the mathematics concepts that they teach (Simon and Blume 1994; Sowder et al. 1998; Thompson and Thompson 1996). Evidence, however, suggests that teachers, particularly at the elementary and middle school levels, often have limited knowledge of the mathematical ideas that are central to the curriculum they are teaching (e.g., Post et al. 1988; Ball 1991; Ma 1999). Therefore, teachers need the opportunity to construct or reconstruct their knowledge of mathematics so that they have a foundation on which to build a practice that requires deep and flexible use of mathematics.

Mathematics experiences for teachers must, however, be different in kind from what is currently available at most institutions of higher education. University mathematics courses were designed for mathematics majors and technical users of mathematics and do not meet the needs of teachers (Howe 1999). According to Ball (1988), "even successful participation in traditional math classes does not necessarily develop the kinds of understanding needed to teach if, as is often the case, success in these classes derives from memorizing formulas and performing procedures" (p. 27).

Professional development must provide teachers with the opportunity to improve their understanding of mathematics content and to reflect critically on their learning experiences. According to *Professional Standards for Teaching Mathematics* (NCTM 1991, p. 128), teachers need opportunities to experience mathematics instruction that will

> enable all learners to experience mathematics as a dynamic engagement in solving problems. These experiences should be designed deliberately to help teachers rethink their conceptions of what mathematics is, what a mathematics class is like, and how mathematics is learned. Instruction should be organized

around searching for solutions to problems and should include continuing opportunities to talk about mathematics...teachers should be encouraged to generalize solutions and communicate results from their exploration of mathematical ideas visually, in writing, or through dialogue and discussion.

Curricular materials designed for students provide one possible source of appropriate mathematical tasks that are likely to engage teachers in worthwhile mathematical activities. The pizza pirate and human chain snapshots in chapter 3 serve as examples of ways in which curricular tasks can be used with teachers. Using curriculum materials that form the core of teachers' instructional programs as the basis for professional development provides a direct link between professional development and what actually happens in classrooms (Russell 1997) and is associated with improved learning outcomes for students (Cohen and Hill 1998).

It is also important to note that experiences can be grounded in mathematics content in less direct ways. For example, by examining students' work (such as the product task and the "extend tiles pattern" task) and by analyzing teachings (such as the case of Fran and Kevin or the videotape of Mrs. Joseph) teachers must begin by making sense of the mathematics. In considering how students solved the problems, teachers must engage with the mathematical ideas that are at the heart of the tasks.

It is widely accepted that most teachers teach the way they were taught. Teachers' experiences as students in mathematics classrooms—where teaching focused on memorization, following given procedures, and coming up with correct answers—often determine their own teaching practices. This "apprenticeship of observation" (Lortie 1975), which begins with early elementary school experiences and often continues through student teaching, has a profound impact on what teachers actually do in their own classrooms.

Professional Development Should Model and Reflect the Pedagogy of Good Instruction

Therefore, professional development experiences need to provide teachers with the opportunity to experience firsthand a form of teaching that facilitates and supports learning—a form that exemplifies the ways in which teachers themselves are being asked to teach. To accomplish this, professional development providers (e.g., teacher educators, staff developers, teacher-leaders) must take to heart the recommendations that teacher education at all levels should model good teaching by posing worthwhile tasks, engaging teachers in discourse, enhancing discourse through the use of a variety of tools, creating learning environments that support and encourage reasoning, and expecting and encouraging teachers to take intellectual risks (NCTM 1991, p. 127).

Each of the snapshots in chapter 3 is intended to exemplify some or all of these characteristics of good instructional practice. Although experiencing good teaching is essential to teachers for reconsidering their current practices, it is not sufficient. Good teaching must also become the object of reflection and discussion so that teachers can begin to identify the features and characteristics of the environment that contributed to and supported their learning. This was evident in the pizza pirate snapshot when Elizabeth Travis asked the teachers to reflect on their experience and how it differed from previous experiences. The preservice teachers were able to identify many aspects of their experience that were different. The identification of these differences may ultimately help teachers create learning environments for their own students that have some of these features.

Professional Development Experiences Should Create Some Disequilibrium for Teachers

Professional development experiences must challenge teachers' current assumptions about what mathematics is, who can do mathematics, and what it means to be successful in mathematics classrooms. According to Ball and Cohen (1999, p. 14) "it would not be sufficient to simply see what one already assumes about students, learning and content; one would also need to see others' assumptions, difference in the content and effects, or unexpected effects of one's own ideas or practices."

Challenging teachers' assumptions will require engaging teachers in new experiences "that cannot be smoothly assimilated with dominant instructional paradigm...[and] are rich enough and flexible enough to engage each of the teachers in accordance with their varying backgrounds and histories" (Schifter and Fosnot 1993, p. 23). By providing teachers with new experiences, and by asking them to reflect on their current practices in light of those experiences, teachers have come to recognize the limitations of their current practices and to begin the process of constructing new ones (Wood, Cobb, and Yackel 1991; Smith 2000a).

Challenging teachers' current assumptions about teaching and learning is likely to cause some discomfort for teachers. As you may recall, the preservice teachers in Elizabeth Travis's class were at times frustrated by their teacher's nondirective approach and experienced some anxiety as they struggled to figure out how to approach and solve the pizza pirate problem. Although experiencing disequilibrium has the potential to stimulate new learning, it can also serve as a rationale for rejecting new ideas. One challenge for teacher educators and other professional development providers will be to find ways to support teachers as they confront and work through these experiences and begin to reconstruct a more reform-oriented practice. Travis accomplished this by asking her students to explain how they were thinking about the problem, by drawing their attention to aspects of it that they may have overlooked, and by steadfastly maintaining her belief that her students could solve the task, given the appropriate support.

Little (1993) calls for professional development that offers "meaning-ful, intellectual, social, and emotional engagement with ideas, with materials, and with colleagues both in and out of teaching" (p. 138). Silver (1996) has suggested that what is needed are communities of collaborative practice where teachers work with colleagues toward shared goals rather than working in isolation.

Professional Development Should Encourage Teacher Collaboration

Despite these calls for more collaboration among teachers, reports suggest that teachers currently spend limited time interacting with their peers (Lubeck 1999), and many teachers express dissatisfaction with this situation (Choy et al. 1993). Providing such opportunities should be seen as a crucial aspect of current reform efforts, since research on school change has found that the extent of interaction between teachers and support providers is related to the degree of change (Fullan 1991). A study of one of the schools participating in the QUASAR project (Stein, Silver, and Smith 1998) highlights the ways in which participating in a community of practice provided sup-port to teachers as they struggled to adopt new practices. Through their collaborative efforts, the QUASAR teachers sustained their growth and development over an extended period of time, supported new mathematics teachers who joined the community, and, in turn, improved their students' learning outcomes.

Several of the snapshots in chapter 3 provide examples of efforts to encourage collaboration among teachers who work in the same school or district. One such example is presented by the teachers at Fair-mont Elementary School, analyzing the snapshot of Mrs. Joseph's teaching. By meeting on an ongoing basis over the course of the semester, the first-, second-, and third-grade teachers had the oppor-tunity to share their experiences in implementing a new curriculum and to engage together in considering new practices. In addition to providing emotional support and practical assistance to teachers as they reconstructed their teaching practices, these meetings also pro-vided an opportunity for teachers to begin to develop a critical atti-tude toward teaching. According to Lord (1994), "collegiality will need to support a critical stance toward teaching. This means more than simply sharing ideas or supporting one's colleagues in the change process. It means confronting traditional practice—the teacher's own and that of his or her colleagues—with an eye toward wholesale revision" (p. 192).

Professional Development Should Take into Account Teachers' Contexts

Professional development cannot be thought of as a "one size fits all" commodity. The context within which teachers work clearly matters. Stein, Smith, and Silver (1999) contend that considering the school context in designing professional development opportunities for teachers is new territory for teacher educators, yet it is critical to the success of change efforts—too little attention to context can derail a reform effort, and too much can paralyze it.

Loucks-Horsley and her colleagues (Loucks-Horsley et al. 1998) discuss several contextual factors that influence professional program design—including students, teachers, and current practices—and urge designers to ground their plans in the realities of the situations in which teachers work. In this view, professional development must take into account the realities of teachers' professional lives—the instructional resources available to them (e.g., texts, manipulatives, calculators), the district and state mandates with which they must comply (e.g., standardized tests, curriculum guidelines), and the constraints imposed by the structure of their school day and year (e.g., forty-two-minute instructional periods, letter grades given quarterly, no planning periods shared with colleagues).

By designing professional development experiences that are sensitive to these realities, reform efforts will have a greater chance of having an impact on the teachers and students who are the target of these efforts. For example, in districts where mandated basic-skills assessments have high stakes for students and teachers, teachers may be reluctant to take time away from skill development to engage students in activities that focus on thinking, reasoning, and problem solving. Instead of ignoring this reality, professional development must help teachers develop practices that will foster students' development of both high- and low-level skills.

For example, in the situation involving the teachers in the Hamlet School District discussed in the human chain snapshot in chapter 3, Charlie Ingram wanted teachers to begin to think about mathematics teaching and learning in a new way. However, he knew that teachers were concerned about the new state standards and how these could be addressed—a critical issue given the context in which the teachers were working. By selecting a mathematically rich task that would engage students in high-level thinking and reasoning while at the same time meeting the goals set forth in the state standards, Charlie was able to address the issue of paramount importance to the teachers and challenge them to think about their current practices.

Another critical but often overlooked aspect of teachers' context is the administrative support for undertaking reform efforts. Mr. Taylor, a teacher participating in the QUASAR project, makes the need for this support clear in the following comment (Brown and Smith 1997, p. 142):

> There were moments when I almost lost control of the class, it got so loud. I think it comes from all these years of being programmed to believe that a good classroom is not loud.... I looked at the door to see if the principal was coming.... I feel uneasy with the noise level. It's something maybe that I have to get used to. But I don't know. Is my principal aware of this?

One way to ensure principals' awareness and support is to provide opportunities for principals and other school administrators to develop a deeper understanding of mathematics education reform and the implications of such reform for the teachers and students in their schools and districts and on their own work as instructional leaders in their buildings. Over the last several years, Nelson and her colleagues have been developing and piloting a curriculum designed specifically to help administrators consider new ideas about mathematics, learning, and teaching (Nelson 1997, 1999). Such efforts will help create school environments where teachers' efforts to teach mathematics in new ways will be both supported and understood.

Loucks-Horsley and her colleagues pose the question "How many professional development efforts have fallen flat, insulting and alienating teachers because they failed to honor their knowledge, skills, culture and experiences?" (1998, p. 176). Teachers possess a wealth of knowledge about their students; the curriculum; and the schools, districts, and communities in which they work. Building on such knowledge is critical to the long-term success of any professional development effort. In their analysis of professional development at Riverside Middle School, Stein, Smith, and Silver (1999) concluded that a failure to acknowledge the expertise of the Riverside teachers led to a standoff between the teachers and the teacher educators with whom they were working and had a long-term negative impact on the collaboration and the reform. As Mr. Hillard, one of the Riverside teachers, commented (Smith and Silver 1998, p. 29):

> We had expertise in one area and ... they had expertise in another. But there was, it seemed as if we were almost polarized with that. And it ended up ... creating a rift that shouldn't have been there.

Although the knowledge that teachers bring to professional development must be acknowledged and appropriately used, teachers cannot be expected to be knowledgeable about all aspects of school reform, subject-matter standards, or professional practice. Collaboration with knowledgeable individuals outside one's own immediate circle is crucial. Outside experts—often university-based educators—bring fresh perspectives and ideas about what has proven successful elsewhere and lend an objective, critical edge to the work (Little 1993).

Little (1993) acknowledges "the difficulty of overcoming long-standing asymmetries in status, power and resources" associated with school-university partnerships, but goes on to note that "as partnerships evolved, they have moved toward greater parity in obligations, opportunities, and rewards" (p. 136). If trusting relationships between practitioners and outside experts are to be established, each party must come to value the expertise that the other brings to the table.

Professional Development Should Make Use of the Knowledge and Expertise of Teachers

47

Professional Development Should Be Sustained and Cohesive

Changing one's teaching calls for a complex transformation of ways of knowing and doing. It requires teachers to change what they know, how they think, and what they do. It takes considerable time and support and should be measured in years rather than days. As Cohen and his colleagues (Cohen et al. 1990) have noted, "changing one's teaching is not like changing one's socks" (p. 163).

To accomplish the kind of deep-seated changes in knowledge, beliefs, and habits of practice that are at the heart of reform efforts, professional development must represent a long-term commitment to teacher growth and development. As such, professional development must consist of a cohesive set of experiences that enhance one another and contribute to a larger, more integrative plan. The "well-intentioned ad-hocism" (Fullan 1991) that has long characterized professional development experiences will not transform teachers' practices.

Professional development must also involve a significant commitment of time. A recent survey completed by teachers during the 1996 administration of the National Assessment of Educational Progress revealed that the majority of students at grades four and eight had teachers who had received fifteen hours or less of professional development in the previous year (Grouws and Smith 2000). Although merely increasing the amount of time available for professional development is unlikely to make a difference, it is equally unlikely that, without being allotted more time, even the best professional development will be effective in accomplishing the ambitious reform agenda. According to *Principals and Standards,* "the work and time of teachers must be structured to allow and support professional development that will benefit them and their students" (NCTM 2000a, p. 19). In a recent position statement on teacher time, the National Council of Teachers of Mathematics (2000b, p. 7) has called for school schedules to be restructured so as to ensure

> the well-planned allocation of personal reflective time for all teachers as well as time for collaborative professional interaction with their peers. This allocated time should include opportunities for deepening understanding of mathematical content; common planning among teachers; the mentoring of new teachers; reading, sharing, and discussing current research and educational literature.

The QUASAR project provides an example of sustained and cohesive support. At each of the six project sites, teachers and administrators from an urban middle school worked in collaboration with university-based teacher educators over a seven-year period to design, implement, and refine reform-oriented mathematics programs that emphasized thinking, reasoning, and problem solving (Silver and Stein 1996; Silver, Smith, and Nelson 1995). These programs were noteworthy for the amount of time they allocated for professional development in general, the amount of time they allowed for teacher colleagues to meet and work together, and their cohesiveness (many of the activities in which teachers engaged were connected to each other in an explicit way). At one of the project sites, for example, teachers met monthly on Saturday mornings to learn about new tools and materials. They also met weekly during their common planning period to prepare lessons and units of instruction that used the materials and tools that had been introduced during the Saturday sessions. As one of the teachers explained (Smith and Brown, forthcoming):

> You get new information and there's this piece in between before you implement it ... that's what we do in those meetings [during common planning time]. We get ready to implement it. We make sure we heard what we heard, and we understood what we heard, then we work it out and then we present it. So it's like I don't see how we could do without any of it.

Teacher professional development begins with university preservice experiences and must continue over the course of a teacher's career. According to Fullan (1991), "good change processes that foster sustained professional development over one's career and lead to student benefits may be one of the few sources of revitalization and satisfaction left for teachers" (p. 131).

Professional development, however, should not remain static over the course of a teacher's career. Professional development programs must continually evaluate what teachers currently know and can do and what they still need to learn and determine the types of experiences that are likely to foster the intended learning. Although many teachers may initially need intensive support for improving their understanding of mathematics content, developing alternative pedagogical strategies, and understanding students better as mathematics learners, it is likely that these needs will change over time.

The teachers at Portsmouth Middle School (Stein, Silver, and Smith, 1998) provide an example of these changing needs. Although their initial support for a new program grew out of their interest in reconstructing their knowledge of mathematics, this support evolved over a three-year period to include a focus on teachers' practice and on the actual work of teaching—developing assessments, refining the curriculum, and communicating with parents. Over time, the teachers became more active members of the broader professional community, making presentations at local, regional, and national conferences, and began conducting professional development for less experienced teachers. Through these new forms of professional development, they continued to examine and refine their own classroom practices.

Professional Development Should Continue over the Course of a Teacher's Career

49

Summary

This chapter illuminates the features of a new form of professional education that fosters the development of teachers over their professional careers. Designing new experiences for teachers that embody these features will require careful attention to many issues. Once a professional development program has been designed, however, it is critically important that professional developers continuously monitor their efforts to ensure that it is working (Loucks-Horsley et al. 1998). In particular, the question "Are we moving toward our goals of improved student learning in mathematics?" (Loucks-Horsley et al. 1998, p. 24)) should become a metric by which all efforts are judged. When the answer to the question is no, planners need to go back to the drawing board and consider ways in which the program could be modified in order to meet these goals. It is important to note that what might work in the first year of a program may need to be revised in subsequent years to meet the changing needs of the teachers and the students.

Although information on whether or not it is working can be collected informally through conversations with participants and observations of teachers in various settings, additional data can also be gathered through more formal and systematic efforts. The next chapter provides suggestions for collecting data that will help assess both short- and long-term impact of professional development on teachers' knowledge, beliefs, and practices and on students' learning.

CHAPTER 5

ASSESSING THE EFFECTIVENESS
OF PROFESSIONAL DEVELOPMENT

Professional development experiences engage teachers in tasks that have the potential to influence classroom practices and take into account the features discussed in the previous chapter. In this chapter, we turn our attention to ways of assessing whether or not the learning opportunities created for teachers have the desired effect on their students.

The effectiveness of professional development should ultimately be measured by the impact that it has on students' learning. Such impact, however, will not occur as the result of teachers' attending one workshop or summer institute. Rather, it will result from a sustained and cohesive effort aimed at helping build teachers' capacity for innovation and intelligent decision making. Looking for or expecting improvements in students' learning in too short a time could result in judging even the best professional education program a failure.

Therefore, in addition to looking for changes in students' performance over time, it is reasonable to consider looking for other signposts of change that are likely to appear as teachers engage in professional education programs. If the goal of these efforts is to change knowledge, beliefs, and habits of practice so as to have an impact on students' learning, then changes in what teachers know, how they think about teaching and learning, and what they do in their classrooms might foreshadow future changes in learning outcomes for students. The suggestions provided in the remainder of this chapter include both long- and short-term ways of measuring impact.

Determining the Impact of a Professional Development Experience on Teachers

Teacher educators should routinely engage in the same type of reflective activity advocated here for teachers. It stands to reason that through reflection on practice, teacher educators too will gain insights into how their actions and interactions in professional development influence teachers' opportunities for learning. Questions such as the following can help teacher educators focus on their work: How did teachers engage in the activity? Who asked the questions, validated answers, provided insights? What did teachers appear to learn from their experiences? How did they learn it? What evidence was there of their learning?

The types of questionnaires that teachers are often asked to complete at the end of a session tend to focus on their satisfaction with the event and provide little useful information about impact. Many teachers, for example, express satisfaction with a session that provided them with "fun" activities to use in the classroom—that met their immediate need to have something to do with students that would be likely to engage them. On the other hand, a session that caused teachers to experience some discomfort, to struggle, or to question some aspect of their practice may not leave them feeling satisfied but may in the long run have a greater impact on their knowledge, beliefs, and habits of practice.

An alternative way to begin to understand how teachers are thinking about the experiences in which they have participated is to ask them to reflect on their experiences in writing at the end of a session or in preparation for a follow-up session. For example, following the discussion of Fran's and Kevin's teaching (see chapter 3 snapshot), teachers were asked to indicate whose classroom they would prefer to be in as a student, and why. This required teachers to begin to clarify their views about what it means to know mathematics, how children learn, and about the type of teaching that will support students' development of mathematical understandings. Additional evidence of the impact of an experience may also manifest itself at some distance from the experience itself. For example, several weeks after the discussion about Fran and Kevin, each teacher was asked to reflect on a videotape of his or her own teaching. One teacher commented that she was "just like Fran" (Stein et al. 2000). She went on to say that she used manipulatives in teaching a lesson, but on reflection, she realized that she used them in a very procedural way, and therefore, like Fran, she limited students' opportunities for learning.

Another way to document impact on teachers' learning is to collect data before and after the professional development experience. The workshop on algebraic thinking for the River City teachers (see the "extend tiles pattern" task snapshot in chapter 3) provides an example of how learning might be assessed. One of the goals of the workshop was to help teachers develop the capacity for critical reflection on teaching. The initial activity of the session involved having teachers watch a short video clip and comment on how the teacher in the video supported or inhibited the students' learning. The clip featured a "reform" classroom that looked very good at first glance—the task was challenging, and the students were engaged. Over the next two days, teachers read and discussed three instructional cases and analyzed students' work on a task closely related to one of them. At the end of the session, the video was shown again, and the same questions were posed. A significantly larger number of inhibiting conditions were identified this time. This result suggested that the experiences in which teachers engaged helped them look beyond the surface—to dig a little deeper as they considered ways in which students' learning could be supported or inhibited. This is not to suggest that one would expect to see immediate changes in these teachers' practice. However, the teachers' postsession responses give evidence that they are beginning to think differently about teaching, and this new thinking can be a solid basis for ongoing work.

Looking for Changes in Teaching

In ongoing professional development programs, teachers could be asked to keep a portfolio of materials documenting their experiences over a specified period of time (e.g., a semester or school year). This might include videotapes of their own teaching, samples of assessments that they have given to students, students' responses to those tasks, and their own reflections on experiences over time. Teachers could be asked periodically to review the materials they have collected and to look for changes in their practices. These reflections could then become the source of discussion among a group of colleagues. Consider, for example, the experience of Elaine Henderson, a teacher who participated in the QUASAR project (Smith 2000a, 2000b). At the end of her first year of participating in the project, Elaine was asked to use the data she had collected to support her perception of her growth over the year. Elaine furnished a video clip from a class that had taken place in October. The tape showed Elaine helping a student measure the perimeter of a pattern train by moving his hands around the outside of the train and asking yes-or-no questions that required little thinking on the part of students. She contrasted this lesson with one she taught near the end of the school year, in which students worked in small groups on a set of challenging problems while she walked around the room asking questions that helped them focus their efforts rather than telling or showing them what to do. For Elaine, the differences in her actions and interactions with students on these two occasions gave evidence that she had changed. Elaine's experience and observations then became a focus of conversation among her colleagues.

Looking at Changes in Mathematical Tasks

Teachers could also be asked to collect a sample of the mathematical tasks used in their classrooms over a specified period of time and to note any changes in the nature of tasks used from the beginning to the end of the time period. This analysis of tasks could include a focus on cognitive demands (i.e., the level of thinking required), communication requirements (i.e., the extent to which explanations and justifications are required), and use of multiple representations and tools. For example, the teachers in the Elmwood School District began collecting tasks that they used in their classes following their initial session with Wendy Carson (see chapter 3 snapshot "Analyzing Cognitive Demands of Mathematical Tasks"). Subsequent meetings focused on the nature of mathematical activity in their classes, with an emphasis on the extent to which students had ongoing opportunities to think, reason, and solve problems. Although Wendy focused on analyzing mathematical tasks using the Mathematical Tasks Framework (Smith and Stein 1998; Stein et al. 2000) there are several other frameworks that could be used for this purpose (e.g., Hiebert et al. 1997; Mokros, Russell, and Economopoulos 1995).

Although using tasks that place higher-level demands on students does not ensure that students are engaging in the tasks at a high level, starting with a high-level task does appear to be a necessary condition for high-level thinking, since low-level tasks almost never result in high-level engagement (Stein and Lane 1996). In addition, although students who participated in classrooms where high-level tasks were used did not consistently engage in the tasks at a high-level, they nevertheless showed greater learning gains on a performance assessment than students did who participated in classrooms where less challenging tasks were used (Stein and Lane 1996).

Looking for Changes in What Students Know and Can Do

A standard way of determining what students have learned or how they are performing is to administer assessments at regular intervals (e.g., beginning and end of a school year; beginning of each school year) and to note the change in students' performance from one time to the next. In order for the information gained from the assessment to be of most value, the assessment selected must be consistent with the instructional program and goals for students' learning. Such assessments could be given at the district level as part of the ongoing evaluation of school programs or by a teacher or group of teachers trying to assess the impact of their instructional efforts within a school building.

At the Classroom Level

In many school districts, the assessments given annually do not align with the instructional goals that target higher-order skills. In such situations, teachers may want to create their own assessments that provide information on students' growth over time on a set of skills that reflect the focus of their instructional efforts. Such assessments could be designed collaboratively by teachers at a particular grade level or within a school building and could be administered in the fall and spring of the school year. Teachers could use the data collected from such assessments to document students' growth over time and reflect on the adequacy both of the instructional program and of students' opportunities to learn the content being assessed. The data could also be used to suggest revisions of the program, and supply information that complements other data available in the system and can amplify reports to administrators and parents about what students know and can do mathematically.

There are many sources of tasks that may be appropriate for this purpose. For example, the regular constructed-response items (e.g., the "extend product pattern" snapshot in chapter 3) and extended constructed-response items (e.g., the "extend tile pattern" snapshot in chapter 3) that have appeared on the National Assessment of Educational Progress may be appropriate for this purpose. Several recent publications contain tasks that have appeared on the assessment, performance data for a national sample of students who took the assessment, and samples of scoring rubrics that were used to score the responses (see Silver and Kenney 2000; Kenney and Silver 1997). In addition, the Balanced Assessment Project has produced a set of books (two at each of three grade levels) that are intended to help teachers implement "meaningful and informative mathematics assessment" (Shannon 1999; Bouck 1999; DeGraw 1999). Each book includes tasks for a particular grade level, at least one sample solution to the task, suggestions for using the task, descriptions of characteristic responses that the task is likely to elicit from students, and rubrics that teachers can use in scoring students' work.

At the District Level

Although many districts continue to administer tests each year that focus primarily on low-level or basic skills, this practice is beginning to change in some places. These changes occur, in large measure, as the result of the efforts by district personnel who are committed to ensuring that testing practices align with new instructional programs that are being adopted and implemented. One school district, for example, recently started using the New Standards Mathematics Reference Exam (NSMRE) (New Standards Project 1996). NSMRE includes extended open-ended and short-answer items that are intended to measure conceptual understanding, mathematical skills and tools, problem solving, reasoning, and mathematical communication. Since this test was aligned with the district's goals for what students should know and be able to do mathematically and with the district's adopted reform curricula (i.e., Everyday Mathematics and Connected Mathematics), the test was considered a good way to assess mathematics learning. The results of the assessment provided a basis for investigating connections between student performance and implementation of the adopted curriculum and for ongoing conversations with teachers in the school district.

Working within the Current System

Over time, many more districts are likely to implement assessments that are targeted at higher-level skills. In the meantime, however, many teachers continue to work in districts where traditional assessments are given each year, and the stakes for students are high. In such places, teachers may be reluctant to move away from traditional practices and curricula for fear that not "teaching to the test" will have a negative impact on their students. The teachers at Riverside Middle School, one of the schools that participated in the QUASAR project, had such a concern (Smith 1997). The test given each year in the school district was not aligned with the new curriculum adopted (with district approval) at Riverside. Since the district test was used to determine placement in the Title I program as well as placement in high school algebra, teachers feared that a drop in performance on the test would have negative implications for their students and for the future of their reform efforts. Although the teacher educators who were working with the teachers tried to assure them that this would not happen, a few adjustments were made to address teachers' concerns. In particular, the sequence of topics in the curriculum was altered so as to ensure that everything that would appear on the spring test would have been taught. In addition, teachers explicitly discussed ways in which they could prepare students for the test without making this preparation the focus of instruction. They selected items from old tests and decided to use them as warm-up activities after a unit in which ideas related to the item were presented. In this way, teachers reasoned, tasks that were procedural in nature would at least be placed in the context of a unit that had focused on an understanding of the underlying concepts.

All the evidence collected at Riverside suggests that students were successful. The district assessment results show that students continued to grow in computational proficiency over the course of the project, even though this was not the focus of the curriculum. For example, the percentage of eighth-grade students demonstrating eligibility for ninth-grade algebra increased dramatically over the years in which Riverside participated in the QUASAR project. Specifically, whereas 8 percent of Riverside eighth graders demonstrated eligibility for algebra in the spring of 1991, 41 percent were eligible for algebra in the spring of 1994. It is interesting to note that there is no evidence that teachers ever actually did anything that could be considered explicit test preparation although such preparation was the focus of a full-day session during the fall of the first year of the project. Riverside students' proficiency with complex mathematical ideas and thinking also substantially improved, as documented by the QUASAR Cognitive Assessment Instrument (Lane 1993). The experience of the Riverside teachers suggests that by focusing on the development and understanding of concepts, rather than on procedures, students are able to apply meaning to more algorithmic situations and hence also to develop proficiency in basic skills.

Summary

This book has made the case that new forms of professional development are needed that can transform teachers' knowledge, beliefs, and habits of practice and ultimately have a positive impact on students' learning outcomes. Although there are professional development programs that have achieved this goal (e.g., Cobb et al. 1991; Carpenter et al. 1989), too often after professional development is conducted, teachers return to their classrooms, and there is no evidence that the teachers and students are changed in any way by the experience. In a recent review of ninety-three studies that examined the effectiveness of various approaches to teacher professional development in mathematics and science, only ten studies included evidence of benefits to students (Kennedy 1999).

Determining the impact of professional development on teachers and on their students is a difficult task, but it is one that must be undertaken. Considerable time, energy, and financial resources are currently being expended on professional development efforts that are not effective. To change this practice, we must continually document and evaluate our efforts so that we more fully understand the programs that make a difference in classroom practices and in the lives of children.

57

REFERENCES

Acquarelli, Kris, and Judith Mumme. "A Renaissance in Mathematics Education Reform." *Phi Delta Kappan* 77 (March 1996), 478–84.

Ball, Deborah Loewenberg. "Prospective Teachers' Understandings of Mathematics: What Do They Bring with Them to Teacher Education?" Paper presented at the annual meeting of the American Educational Research Association, New Orleans, La., April 1988.

———. "Research on Teaching Mathematics: Making Subject Matter Knowledge Part of the Equation." In *Advances in Research on Teaching*, vol. 2, *Teacher's Knowledge of Subject Matter as It Relates to Their Teaching Practices*, edited by Jere Brophy, pp. 1–48. Greenwich, Conn.: JAI Press, 1991.

———. "With an Eye on the Mathematical Horizon: Dilemmas of Teaching Elementary School Mathematics." *Elementary School Journal* 93, no. 4 (March 1993), 373–97.

———. "Teacher Learning and the Mathematics Reforms: What We Think We Know and What We Need to Learn." *Phi Delta Kappan* 77 (March 1996), 500–508.

———. "A Practice-Based Approach to Teacher Education: The Potential Affordance and Difficulties." Paper presented at the annual meeting of the Association of Mathematics Teacher Educators, Costa Mesa, Calif., January 2001.

Ball, Deborah Loewenberg, and David K. Cohen. "Developing Practice, Developing Practitioners: Toward a Practice-Based Theory of Professional Education." In *Teaching as the Learning Profession: Handbook of Policy and Practice*, edited by Linda Darling-Hammond and Gary Sykes, pp. 3–32. San Francisco: Jossey-Bass, 1999.

Barnett, Carne. "Building a Case-Based Curriculum to Enhance the Pedagogical Content Knowledge of Mathematics Teachers." *Journal of Teacher Education* 42 (September/October 1991), 263–72.

Barnett, Carne, Donna Goldenstein, and Babette Jackson, eds. *Fractions, Decimals, Ratios, and Percents: Hard to Teach and Hard to Learn?* Portsmouth, N.H.: Heinemann, 1994a.

———. *Fractions, Decimals, Ratios, and Percents: Hard to Teach and Hard to Learn? Facilitator's Discussion Guide.* Portsmouth, N.H.: Heinemann, 1994b.

Blume, Glendon W., and David S. Heckman. "What Do Students Know about Algebra and Functions?" In *Results from the Sixth Mathematics Assessment of the National Assessment of Educational Progress*, edited by Patricia Ann Kenney and Edward A. Silver, pp. 225–78. Reston, Va.: National Council of Teachers of Mathematics, 1997.

Bouck, Mary, ed. *Middle Grades Assessment: Balanced Assessment for the Mathematics Curriculum*. White Plains, N.Y.: Dale Seymour Publications, 1999.

Brown, Catherine A., and Margaret S. Smith. "Supporting the Development of Mathematical Pedagogy." *Mathematics Teacher* 90 (February 1997), 138–43.

Carnegie Learning. *Cognitive Tutor Algebra I.* Pittsburgh, Pa.: Carnegie Learning, 1998, 1999.

Carpenter, Thomas P., Elizabeth Fennema, Penelope L. Peterson, Chi-Pang Chiang, and Megan Loef. "Using Knowledge of Children's Mathematics Thinking in Classroom Teaching: An Experimental Study." *American Educational Research Journal* 26, no. 4 (winter 1989), 499–531.

Choy, Susan P., Sharon A. Bobbitt, Robin R. Henke, Elliott A. Medrich, L. J. Horn, and J. Lieberman. *America's Teachers: Profiles of a Profession.* Washington, D.C.: National Center for Education Statistics, U. S. Department of Education, 1993.

Cobb, Paul, Terry Wood, Erna Yackel, John Nicholls, Grayson Wheatley, Beatriz Trigatti, and Marcella Perlwitz. "Assessment of a Problem-Centered Second-Grade Mathematics Project." *Journal for Research in Mathematics Education* 22 (January 1991), 3–29.

Cohen, David, Penelope Peterson, Suzanne Wilson, R. Wheaten, Janine Remillard, and N. Weimers. *Effects of State-Level Reform of Elementary School Mathematics Curriculum on Classroom Practice.* Elementary Subjects Center Series, no. 25. East Lansing, Mich.: Michigan State University, Center for the Learning and Teaching of Elementary Subjects and National Center for Research on Teacher Education, 1990.

Cohen, David K., and Heather C. Hill. *State Policy and Classroom Performance: Mathematics Reform in California.* Consortium for Policy Research in Education (CPRE) Policy Briefs, Research Briefs, no. 23. Philadelphia: CPRE, 1998.

Connected Mathematics Project and Michigan State University. *Connected Mathematics.* Menlo Park, Calif.: Dale Seymour Publications, 1996.

Darling-Hammond, Linda. "Reframing the School Reform Agenda: Developing Capacity for School Transformation." *Phi Delta Kappan* 74 (June 1993): 752–61.

Darling-Hammond, Linda, and Gary Sykes, eds. *Teaching as the Learning Profession: Handbook of Policy and Practice.* San Francisco: Jossey-Bass, 1999.

Darling-Hammond, Linda, and Milbrey W. McLaughlin. "Policies That Support Professional Development in an Era of Reform." *Phi Delta Kappan* 76 (April 1995): 597–604.

DeGraw, Mishaa, ed. *Elementary Grades Assessment: Balanced Assessment for the Mathematics Curriculum.* White Plains, N.Y.: Dale Seymour Publications, 1999.

Doyle, Walter. "Work in Mathematics Classes: The Context of Students' Thinking during Instruction." *Educational Psychologist* 23 (February 1988), 167–80.

Evans, Christine Sobrary. "When Teachers Look at Student Work." *Educational Leadership* 50 (February 1993), 71–72.

Foreman, Linda, and Albert Bennett. *Visual Mathematics Course.* Salem, Oreg.: Math Learning Center, 1996.

Friel, Susan N., and Lisa W. Carboni. "Using Video-Based Pedagogy in an Elementary Mathematics Methods Course." *School Science and Mathematics* 100 (March 2000), 118–27.

Fullan, Michael. *The New Meaning of Educational Change.* New York: Teachers College Press, 1991.

Ginsburg, Herbert P., Susan F. Jacobs, and Luz Stella Lopez. *The Teacher's Guide to Flexible Interviewing in the Classroom: Learning What Children Know about Math.* Needham Heights, Mass.: Allyn & Bacon, 1998.

Grouws, Douglas A., and Margaret Schwan Smith. "Findings from NAEP on the Preparation and Practices of Mathematics Teachers." In *Results from the Seventh Mathematics Assessment of the National Assessment of Educational Progress*, edited by Edward A. Silver and Patricia Ann Kenney, pp. 107–39. Reston, Va.: National Council of Teachers of Mathematics, 2000.

59

Hiebert, James, Thomas P. Carpenter, Elizabeth Fennema, Karen C. Fuson, Diana Wearne, Hanlie Murray, Alwyn Olivier, and Piet Human. *Making Sense: Teaching and Learning Mathematics with Understanding.* Portsmouth, N.H.: Heinemann, 1997.

Howe, Roger, review of *Knowing and Teaching Elementary Mathematics: Teachers' Understanding of Fundamental Mathematics in China and the United States,* by Liping Ma, *Journal for Research in Mathematics Education* 30 (November 1999) 579–89.

Huberman, Michael. "The Model of an Independent Artisan in Teachers' Professional Relations." In *Teachers' Work: Individuals, Colleagues, and Contexts,* edited by Judith W. Little and Milbrey W. McLaughlin. New York: Teachers College Press, 1993.

Kamii, Constance. *Double-Column Addition: A Teacher Uses Piaget's Theory.* New York: Teachers College Press, 1989. Video cassette.

Kazemi, Elham, and Megan Loef Franke. "Understanding Teacher Learning as Changing Participation in Communities of Practice." In *Proceedings of the 22nd Annual Meeting of the North American Chapter of the International Group for the Psychology of Mathematics Education,* pp. 561–66. Columbus, Ohio: ERIC Clearinghouse for Science, Mathematics, and Environmental Education, 2000.

Kennedy, Mary M. "Form and Substance in Mathematics and Science Professional Development." *National Institute for Science Education Brief* 3 (November 1999): 1–7.

Kenney, Patricia Ann, and Edward A. Silver. "Probing the Foundations of Algebra: Grade-4 Pattern Items in NAEP." *Teaching Children Mathematics* 3 (February 1997a): 268–74.

Kenney, Patricia Ann, and Edward A. Silver, eds. *Results from the Sixth Mathematics Assessment of the National Assessment of Educational Progress.* Reston, Va.: National Council of Teachers of Mathematics, 1997.

Lamon, Susan J. *Teaching Fractions and Ratios for Understanding.* Mahwah, N.J.: Lawrence Erlbaum Associates, 1999.

Lane, Suzanne. "The Conceptual Framework for the Development of a Mathematics Performance Assessment." *Educational Measurement: Issues and Practice* 14, no. 1 (summer 1993): 16–23.

Little, Judith Warren. "Teachers' Professional Development in a Climate of Educational Reform." *Educational Evaluation and Policy Analysis* 15, no. 2 (1993): 129–51.

Lord, Brian. "Teachers' Professional Development: Critical Colleagueship and the Role of Professional Communities." In *The Future of Education Perspectives on National Standards in America,* edited by Nina Cobb, pp. 175–204. New York: College Board, 1994.

Lortie, Dan C. *Schoolteacher: A Sociological Study.* Chicago: University of Chicago Press, 1975.

Loucks-Horsley, Susan, Peter Hewson, Nancy Love, and Katherine E. Stiles. *Designing Professional Development for Teachers of Science and Mathematics.* Thousand Oaks, Calif.: Corwin Press, 1998.

Lubeck, Sally. "Teachers and the Teaching Profession in the United States." In *The Educational System in the United States: Case Study Findings,* pp. 173–214. Washington, D.C.: U.S. Department of Education, 1999.

Ma, Liping. *Knowing and Teaching Elementary Mathematics: Teachers' Understanding of Fundamental Mathematics in China and the United States.* Hillsdale, N.J.: Lawrence Erlbaum Associates, 1999.

Merseth, Katherine K., and Catherine A. Lacey. "Weaving Stronger Fabric: The Pedagogical Promise of Hypermedia and Case Methods in Teacher Education." *Teaching and Teacher Education* 9 (June 1993): 283–99.

Mokros, Jan, Susan Jo Russell, and Karen Economopoulos. *Beyond Arithmetic: Changing Mathematics in the Elementary Classroom.* Palo Alto, Calif.: Dale Seymour Publications, 1995.

Moon, Jean. *Developing Judgment: Assessing Children's Work in Mathematics.* Portsmouth, N.H.: Heinemann, 1997.

National Board for Professional Teaching Standards (NBPTS). *Middle Childhood and Early Adolescence/Mathematics: Standards for National Board Certification.* Washington, D.C.: NBPTS, 1997.

National Council of Teachers of Mathematics (NCTM). *Curriculum and Evaluation Standards for School Mathematics.* Reston, Va.: NCTM, 1989.

———. *Professional Standards for Teaching Mathematics.* Reston, Va.: NCTM, 1991.

———. *Assessment Standards for School Mathematics.* Reston, Va.: NCTM, 1995.

———. *Principles and Standards for School Mathematics.* Reston, Va.: NCTM, 2000a.

———. "NCTM Position on Teacher Time." *News Bulletin* 36 (May/June 2000b): 7.

Nelson, Barbara Scott. "Lenses on Learning: How Administrators' Ideas about Mathematics, Learning, and Teaching Influence Their Approaches to Action in an Era of Reform." *Journal of Mathematics Teacher Education* 1, no. 2 (1997): 191–215.

———. *Building New Knowledge by Thinking: How Administrators Can Learn What They Need to Know about Mathematics Education Reform.* Newton, Mass.: Center for the Development of Teaching, Education Development Center, 1999.

New Standards Project. *New Standards Reference Examinations: Mathematics.* San Antonio, Tex.: Harcourt Brace, 1996.

Parker, Diane L., and Anthony J. Picard. "Portraits of Susie: Matching Curriculum, Instruction, and Assessment." *Teaching Children Mathematics* 3 (March 1997): 376–82.

Post, Thomas, Guershon Harel, Merlyn Behr, and Richard Lesh. "Intermediate Teachers' Knowledge of Rational Number Concepts." In *Integrating Research on Teaching and Learning Mathematics,* edited by Elizabeth Fennema, Thomas P. Carpenter, and Susan J. Lamon, pp. 177–98. Madison, Wis.: Center for Education Research, 1988.

Russell, Susan Jo. "The Role of Curriculum in Teacher Development." In *Reflecting on Our Work: NSF Teachers Enhancement in K–6 Mathematics,* edited by Susan N. Friel and George W. Bright, pp. 247–54). Lanham, Md.: University Press of America, 1997.

Schifter, Deborah, and Catherine T. Fosnot. *Reconstructing Mathematics Education: Stories of Teachers Meeting the Challenge of Reform.* New York: Teachers College Press, 1993.

Schifter, Deborah, ed. *What's Happening in Math Class? Envisioning New Practices through Teacher Narratives.* New York: Teachers College Press, 1996.

———. *What's Happening in Math Class? Reconstructing Professional Identities.* Vol. 2. New York: Teachers College Press, 1996.

Shannon, Ann, ed. *High School Assessment: Balanced Assessment for the Mathematics Curriculum*. White Plains, N.Y.: Dale Seymour Publications, 1999.

Shulman, Lee S. "Those Who Understand: Knowledge Growth in Teaching." *Educational Researcher* 15 (February 1986), 4–14.

———. "Toward a Pedagogy of Cases." In *Case Methods in Teacher Education*, edited by Judith H. Shulman, pp. 1–30. New York: Teachers College Press, 1992.

Silver, Edward A. "Mathematical Thinking and Reasoning for All Students: Moving from Rhetoric to Reality." In *Selected Lectures from the 7th International Congress on Mathematics Education*, edited by David F. Robitaille, David H. Wheeler, and Carolyn Kieran, pp. 311–25. Quebec, Canada: Les Presses de l'Université Laval, 1994.

———. "Moving beyond Learning Alone and in Silence: Observations from the QUASAR Project concerning Some Challenges and Possibilities of Communication in Mathematics Classrooms." In *Innovations in Learning: New Environments for Education*, edited by Leona Schauble and Robert Glaser, pp. 289–325. Mahwah, N.J.: Lawrence Erlbaum Associates, 1996.

Silver, Edward A., and Mary Kay Stein. "The QUASAR Project: The 'Revolution of the Possible' in Mathematics Instructional Reform in Urban Middle Schools." *Urban Education* 30 (January 1996): 476–521.

Silver, Edward A., and Patricia Ann Kenney, eds. *Results from the Seventh Mathematics Assessment of the National Assessment of Educational Progress*. Reston, Va.: National Council of Teachers of Mathematics, 2000.

Silver, Edward A., Cengiz Alacaci, and Despina A. Stylianou. "Students' Performance on Extended Constructed-Response Tasks." In *Results from the Seventh Mathematics Assessment of the National Assessment of Educational Progress*, edited by Edward A. Silver and Patricia Ann Kenney, pp. 301–41. Reston, Va.: National Council of Teachers of Mathematics, 2000.

Silver, Edward, Margaret S. Smith, and Barbara Nelson. "The QUASAR Project: Equity Concerns Meeting Mathematics Education Reform in the Middle School." In *New Directions for Equity in Mathematics Education*, edited by Walter Secada, Elizabeth Fennema, and Lisa Adajian, pp. 9–56. New York: Cambridge University Press, 1995.

Simon, Martin A., and Deborah Schifter. "Towards a Constructivist Perspective: An Intervention Study of Mathematics Teachers." *Educational Studies in Mathematics* 22, no. 4 (August 1991): 309–31.

Simon, Martin A., and Glendon W. Blume. "Building and Understanding Multiplicative Relationships: A Study of Prospective Elementary Teachers." *Journal for Research in Mathematics Education* 25 (November 1994): 472–94.

Smith, Margaret Schwan. "Riverside Middle School: School Reform Supported by an Innovative Curriculum." Paper presented as part of a symposium, "Improving Mathematics Instruction in Urban Middle Schools: Facilitating and Inhibiting Conditions Encountered in the QUASAR Project," at the annual meeting of the American Educational Research Association, Chicago, March 1997.

———. "An Examination of the Impact of Using an Innovative Mathematics Curriculum for Students as a Resource for the Professional Development of Teachers." Paper presented at the annual meeting of the American Educational Research Association, Montreal, April 1999.

———. "Balancing Old and New: An Experienced Middle School Teacher's Learning in the Context of Mathematics Instructional Reform." *Elementary School Journal* 100 (March 2000a): 351–75.

———. "Redefining Success in Mathematics Teaching and Learning." *Mathematics Teaching in the Middle School* 5 (February 2000b): 378–82, 386.

———. "Using Cases to Discuss the Changing Role of the Mathematics Teacher." *Mathematics Teaching in the Middle School* (in press).

Smith, Margaret Schwan, and Catherine A. Brown. "Teacher Professional Development and Support." In *Teaching Mathematics for a Change: Evidence from the QUASAR Project Regarding the Challenges and Possibilities of Instructional Reform in Urban Middle Schools.* Forthcoming.

Smith, Margaret Schwan, and Edward A. Silver. "Professional Development at Riverside Middle School." Paper presented as part of a symposium, "Dilemmas in Designing Professional Development for Science and Mathematics Teachers," at the annual meeting of the American Educational Research Association, San Diego, April 1998.

Smith, Margaret Schwan, and Mary Kay Stein. "Selecting and Creating Mathematical Tasks: From Research to Practice." *Teaching Mathematics in the Middle School* 3 (February 1998): 344–50.

Sowder, Judith, Barbara Armstrong, Susan Lamon, Martin Simon, Larry Sowder, and Alba Thompson. "Educating Teachers to Teach Multiplicative Structures in the Middle Grades." *Journal of Mathematics Teacher Education* 1, no. 2 (1998): 127–55.

Sparks, Dennis, and Susan Loucks-Horsley. "Models of Staff Development." In *Handbook of Research on Teacher Education*, edited by W. Robert Houston, Martin Haberman, and John Sikula, pp. 234–50. New York: Macmillan, 1990.

Stein, Mary Kay, and Jane W. Bovalino. "Manipulatives: One Piece of the Puzzle." *Mathematics Teaching in the Middle School* 6 (February 2001): 356–59.

Stein, Mary Kay, and Suzanne Lane. "Instructional Tasks and the Development of Student Capacity to Think and Reason: An Analysis of the Relationship between Teaching and Learning in a Reform Mathematics Project." *Educational Research and Evaluation* 2 (October 1996): 50–80.

Stein, Mary Kay, Edward A. Silver, and Margaret Schwan Smith. "Mathematics Reform and Teacher Development from the Community of Practice Perspective." In *Thinking Practices: A Symposium on Mathematics and Science Learning*, edited by James G. Greeno and Shelley Goldman, pp. 17–52. Hillsdale, N.J.: Lawrence Erlbaum Associates, 1998.

Stein, Mary Kay, Margaret Schwan Smith, and Edward A. Silver. "The Development of Professional Developers." *Harvard Educational Review* 69, no. 3 (fall 1999): 237–69.

Stein, Mary Kay, Margaret S. Smith, Marjorie A. Henningsen, and Edward A. Silver. *Exploring Cognitively Challenging Mathematical Tasks: A Casebook for Teacher Professional Development.* New York: Teacher's College Press, 2000.

Stigler, James W., and James Hiebert. *The Teaching Gap.* New York: Free Press, 1999.

Stylianou, Despina, and Margaret Schwan Smith. "Examining Student Responses: A Strategy for Developing Pre-Service Elementary Teachers' Understanding of Algebra." In *Algebra across the Grades*, 2000 Yearbook of the Pennsylvania Council of Teachers of Mathematics, edited by M. Kathleen Heid, Margaret Schwan Smith, and Glendon W. Blume, pp. 23–32. N.p.: Pennsylvania Council of Teachers of Mathematics, 2000.

Thompson, Alba G., and Patrick W. Thompson. "Talking about Rates Conceptually, Part II: Mathematical Knowledge for Teaching." *Journal for Research in Mathematics Education* 27 (January 1996): 2–24.

Thompson, Charles L., and John S. Zeuli. "The Frame and the Tapestry: Standards-Based Reform and Professional Development." In *Teaching as the Learning Profession: Handbook of Policy and Practice*, edited by Linda Darling-Hammond and Gary Sykes, pp. 341–75. San Francisco: Jossey-Bass, 1999.

Wilcox, Sandra K., Pamela Schram, Glenda Lappan, and Perry Lanier. "The Role of a Learning Community in Changing Preservice Teachers' Knowledge and Beliefs about Mathematics Education." Paper presented at the annual meeting of the American Educational Research Association, Boston, April 1991.

Wood, Terry, Paul Cobb, and Erna Yackel. "Change in Teaching Mathematics: A Case Study." *American Educational Research Journal* 28, no. 3 (May 1991): 587–616.

Zullie, Matthew E. *Fractions with Pattern Blocks*. Chicago: Creative Publications, 1988.

APPENDIX A

INNOVATIVE CURRICULA

The National Science Foundation (NSF) funded thirteen curriculum projects that are aligned with the NCTM *Standards* (1989, 1991, 2000a) and intended to support reform-oriented instruction. Each of these curricula contains a storehouse of good mathematical tasks that could be used in a range of professional development settings.

NSF-Funded Elementary Curricula

Curriculum	Developer	Publisher
Investigations in Number, Data, and Space	TERC 2067 Massachusetts Avenue Cambridge, MA 02140 Phone: (617) 547-0430 Contact: Lorraine Brooks E-mail: lorraine_brooks@terc.edu Web: www.terc.edu/investigations/index.html	Scott Foresman Contact: Linda Dodge Product Manager, Mathematics 16 Linden Avenue Greenfield, MA 01301 Phone: (413) 773-0434 Fax: (413) 772-0780 E-mail: linda.dodge@scottforesman.com Web: www.cuisenaire-dsp.com/start_full.html
Math Trailblazers	TIMS Project Institute for Math and Science Education University of Illinois at Chicago 2075 SEL M/C 250 950 South Halsted Street Chicago, IL 60607 Contact: Joan Bieler or Cathy Kelso Phone: (800) 454-TIMS E-mail: jbieler@uic.edu or ckelso@uic.edu Web: www.math.uic.edu/IMSE/MTB/mtb.html	Kendall/Hunt Publishing Company Contact: Dennis Jaeger 4050 Westmark Drive Dubuque, IA 52004-1840 Phone: (800) 542-6657 Fax: (319) 589-1071 E-mail: khelhi@aol.com Web: www.kendallhunt.com
Everyday Mathematics	University of Chicago 5835 South Kimbark Avenue Chicago, IL 60637 Contact: Andy Isaacs Phone: (773) 702-9639 E-mail: aisaacs@midway.uchicago.edu	Everyday Learning Corporation Contact: Toni Fleming Vice President, Marketing and Teacher Education 2 Prudential Plaza, Suite 1200 Chicago, IL 60601 Phone: (800) 382-7670 Fax: (312) 540-5848 E-mail: tfleming@tribune.com Web: www.everydaylearning.com /Pages/everyday.html

65

NSF-Funded Middle School Curricula

Curriculum	Developer	Publisher
Connected Mathematics	Michigan State University Contact: Betty Phillips E-mail: cmp@math.msu.edu Web: www.mth.msu.edu/cmp	Prentice Hall Phone: (800) 872-1100 Web: www.phschool.com
MathScape	Education Development Center Contact: Susan Janssen E-mail: mathscape@edc.org Web: www.edc.org/mathscape	Creative Publications Phone: (800) 624-0822 Web: www.creativepublications.com
Math Thematics	University of Montana Contact: Rick Billstein E-mail: rickb@selway.umt.edu Web: www.math.umt.edu/~stem	McDougal Littell Phone: (800) 323-4068 (ext. 3349) Web: www.mlmath.com/mathmtcs.htm
Mathematics in Context	University of Wisconsin—Madison Contact: Meg Meyer E-mail: mrmeyer2@facstaff.wisc.edu	Encyclopaedia Britannica Phone: (800) 554-9862 Web: www.ebmic.com
Pathways to Algebra and Geometry	MMAP/Pathways Implementation Center Contact: Jennifer Knudsen E-mail: jknudse@wested.org Web: mmap.wested.org	Voyager Expanded Learning Web: www.iamvoyager.com

NSF-Funded Secondary Curricula

Curriculum	Developer	Publisher
Mathematics: Modeling Our World	COMAP, Inc. 57 Bedford Street, Suite 210 Lexington, MA 02173 Contact: Sol Garfunkel Toll-Free: (800) 772-6627 Phone: (781) 862-7878 Fax: (781) 863-1202 E-mail: sol@pop.comap.com Web: www.comap.com/highschool /projects/arise.htm	W.H. Freeman and Company 345 Park Avenue South New York, NY 10010 Phone: (800) 446-8923 Web: whfreeman.com/highschool/book
Contemporary Mathematics in Context	Core-Plus Mathematics Project (CPMP) Contact: Beth Ritsema or Marcia Weinhold Department of Mathematics and Statistics Western Michigan University Kalamazoo, MI 49008 Phone: (616) 387-4562 Fax: (616) 387-4530 E-mail: cpmp@wmich.edu Web: www.wmich.edu/cpmp	Everyday Learning Corporation Two Prudential Plaza Suite 1175 Chicago, IL 60601 Phone: (800) 322-6284 Web: www.everydaylearning.com /Pages/contemp.html
Math Connections	Connecticut Business and Industry Association (CBIA) Education Foundation Contact: June Ellis Phone: (860) 721-7010 E-mail: mathconx@aol.com Web: www.mathconnections.com	It's About Time 84 Business Park Drive Armonk, NY 10504 Phone: (888) 698-TIME E-mail: itstimefor@aol.com
Integrated Mathematics: A Modeling Approach Using Technology	SIMMS Integrated Mathematics Dissemination Center 401 Linfield Hall Montana State University Bozeman, MT 59717-2810 Contact: Gary Bauer Phone: (406) 994-7066 or (800) 693-4060 Fax: (406) 994-3733 E-mail: gbauer@math.montana.edu Web: www.montana.edu/wwwsimms	Pearson Custom Publishing Contact: Lynette Felix, Sales Rep. SIMMS Integrated Mathematics Project 401 Linfield Hall Bozeman, MT 59717 Phone: (800) 693-4060 E-mail: felix@math.montana.edu Purchase orders can be faxed to (406) 994-3733 Web: www.montana.edu/wwwsimms
Interactive Mathematics Programs	IMP Contact: Janice Bussey Phone: (888) MATH-IMP or (415) 332-3328 E-mail: IMP@math.sfsu.edu Web: www.mathimp.org	Key Curriculum Press 1150 65th Street Emeryville, CA 94608 Phone: (800) 995-MATH Fax: (800) 541-2442 Web: www.keypress.com/index.html

CURRICULUM IMPLEMENTATION CENTERS

NSF also funded four centers that are intended to support dissemination and implementation of the curricula and are sources of information about the curricula. In addition, the centers provide sample lessons from the curricula that can be downloaded.

Focus of Implementation	Publisher
K–12	**K–12 Mathematics Curriculum Center** Education Development Center, Inc. 55 Chapel Street Newton, MA 02458-1060 Phone: (800) 332-2429 E-mail: mcc@edc.org Web: www.edu.org/mcc
Elementary	**The ARC Center** COMAP, Inc. 57 Bedford Street, Suite 210 Lexington, MA 02173 Phone: (800) 772-6627 (ext. 50) E-mail: arccenter@mail.comap.com Web: www.arccenter.comap.com
Middle	**The Show-Me Center** University of Missouri 104 Stewart Hall Columbia, MO 65211 Phone: (573) 884-2099 E-mail: center@showme.missouri.edu Web: www.showmecenter.missouri.edu
Secondary	**COMPASS** Ithaca College 306 Williams Hall Ithaca, NY 14850 Phone: (800) 688-1829 E-mail: compass@ithaca.edu Web: www.ithaca.edu/compass

ADDITIONAL CURRICULAR RESOURCES

Several other innovative curricula may also serve as sources of interesting mathematical tasks.

Math Alive (Visual Mathematics)

This is a complete three-year curriculum appropriate for grades 5–8. An important feature of the curriculum is the use of models, manipulatives, sketches, and diagrams to support students' development of a conceptual understanding of mathematics and create mental images that help them retain and recall this information.

The Math Learning Center Materials
P.O. Box 3226
1850 Oxford Street SE
Salem, OR 97302
Phone: (800) 575-8130
Web: www.mlc.pdx.edu

Cognitive Tutor Algebra

This is a first-year algebra course that integrates technology in its instructional design. An important feature of the program is an intelligent tutor that provides each students with an individualized coach. This program been designated an exemplary program by the U.S. Department of Education.

Carnegie Learning, Inc.
372 N. Craig Street, Suite 101
Pittsburgh, PA 15213
Phone: (412) 683-6284
Web: www.carnegielearning.com

Concepts in Algebra: A Technological Approach

This is a one-year course that focuses on developing students' quantitative understanding of functions in a variety of real-world problem situations. The use of technological tools is integrated throughout the curriculum.

Everyday Learning Corporation
2 Prudential Plaza, Suite 1200
Chicago, IL 60601
Phone: (800) 382-7670
Web: www.everydaylearning.com

APPENDIX B

RESOURCES FOR PROFESSIONAL DEVELOPMENT

TE-MAT (Teacher Education Materials) Database

The TE-MAT (Teacher Education Materials) project was funded through a National Science Foundation grant to Horizon Research, Inc., to develop an online resource for grades K–12 mathematics and science professional development providers. The Web site is intended to increase the accessibility of professional development materials and to encourage their effective use in preservice and in-service programs. The site includes an underlying conceptual framework that highlights critical areas to be considered in the design of effective preservice and in-service programs, a searchable database of reviews of materials, and practitioner essays in which professional development providers share insights about how they have incorporated specific materials into their programs. On completion of the project in March 2002, the database of reviews of materials will be maintained by the Eisenhower National Clearinghouse as part of their ongoing dissemination of work in science and mathematics education. The TE-MAT Web site is found at www.te-mat.org.

71

The resources listed below represent a subset of the practice-based materials that could be drawn on in designing or conducting the type of professional development described in this book—professional development that uses practice-based materials as the basis for professional learning tasks. The list is not intended to be exhaustive but rather to provide the reader with a set of resources that can get them started. (Those followed by an asterisk can also be found in the TE-MAT database described on the previous page.)

Barnett, Carne, Donna Goldenstein, and Babette Jackson, eds. *Fractions, Decimals, Ratios, and Percents: Hard to Teach and Hard to Learn?* Portsmouth, N.H.: Heinemann, 1994.*

The twenty-nine cases in this book portray the dilemmas encountered in teaching fractions, decimals, ratios, and percents in actual fourth-through eighth-grade classrooms. Each case includes examples of authentic students' work and references to related research. The cases provide a context for focusing on mathematical content, understanding children's thinking, and examining the decisions teachers make in actual classroom situations. The facilitation guide provides teacher educators with information on using cases in general and on guiding the discussions of each specific case in the book.

Barnett, Carne, and Pam Tyson. *Enhancing Mathematics Teaching through Case Discussion.* San Francisco: WestEd, 1994. (For ordering information, call 415-565-3021.)

In this video, teachers engage in a case discussion focused on children's thinking, mathematics, language issues, and teaching. The video provides teacher educators with a model of how case discussions can spark new ideas and challenge old beliefs while simultaneously providing support and encouragement for teachers to make changes in their thinking.

Blume, Glendon W., Judith S. Zawojewski, Edward A. Silver, and Patricia A. Kenney. "Focusing on Worthwhile Mathematical Tasks in Professional Development: Using a Task from the National Assessment of Educational Progress." *Mathematics Teacher* 91 (February 1998): 156-61.

This article shows how a worthwhile mathematical task can generate rich activities for professional development. Featured in the article is a released constructed-response item from the 1992 Twelfth-Grade NAEP that can be solved through many different strategies (tables, graphs, symbolic notation, pictures, or deductive reasoning). By solving the task in a variety of ways and comparing the different approaches, the authors note that teachers engaged in interesting and instructive discussions about students' thinking, assessment, and the validity of non-symbolic approaches. The article closes by referring to other sources of NAEP tasks and by providing suggestions on how such tasks might be used as the basis of professional development activities.

Bryant, Deborah, and Mark Driscoll. *Exploring Classroom Assessment in Mathematics: A Guide for Professional Development.* Reston, Va.: National Council of Teachers of Mathematics, 1998.*

This book provides staff developers with a resource to assist K–12 mathematics teachers in changing their assessment practices to reflect current principles of mathematics education reform. Central to the guidebook are six teacher investigations, two- to three-hour professional development workshops (entitled "Experiencing a Task," "Observing Problem Solving," "Examining Student Work," "Developing Tasks," "Developing a Rubric," and "Planning Assessment") that can be used independently or as a coherent, ongoing professional development program. The investigations also aim to challenge teachers' notions about the teaching and learning of mathematics and to push teachers to continually consider and seek evidence of student learning.

Burns, Marilyn. *Mathematics for Middle School: Grades 6–8*, Parts 1, 2, and 3. New Rochelle, N.Y.: Cuisenaire, 1989.*

The videos in this series demonstrate exemplary teaching in middle school mathematics classrooms. Each part of the series contains a twenty-minute video for viewing by teacher participants and a teacher development guide to provide teacher educators with facilitation support. Although each video has a slightly different focus (Part 1, the student's role; Part 2, the teacher's role; Part 3, the role of the curriculum), they all portray students engaged in problem-solving activities covering a wide range of important mathematical topics for middle school.

Carpenter, Thomas, Elizabeth Fennema, Megan Loef Franke, Linda Levi, and Susan B. Empson. *Children's Mathematics: Cognitively Guided Instruction.* Portsmouth, N.H.: Heinemann, 1999.*

This book is intended to assist teachers of elementary school mathematics in understanding children's intuitive mathematical thinking and in applying this knowledge to plan instruction. The book is accompanied by two CDs. The first addresses children's strategies for computation with single-digit and multidigit numbers and includes clips of actual classrooms where children are working on problems discussed in the book. The second offers suggestions for implementing and supporting cognitively guided instruction in the classroom. The book is written for teachers and would be well suited for a study group or ongoing workshop of elementary teachers.

Clarke, David. *Constructive Assessment in Mathematics: Practical Steps for Classroom Teachers.* Emeryville, Calif.: Key Curriculum Press, 1997.*

This book is intended to help practicing or prospective teachers apply current research ideas on assessment in their own classrooms. The author provides a rationale for reforming assessment practices to consist of much more than assigning a letter grade. As a professional development resource, this book is both informative and capable of stimulating interesting discussion.

Driscoll, Mark, and Deborah Bryant. *Learning about Assessment, Learning through Assessment.* Washington, D.C.: Mathematical Sciences Education Board, 1988.*

This booklet is intended to guide staff developers as they assist K–12 mathematics teachers in learning about assessment and how to use it as a stimulus for instructional change. The authors refer to current research, their own experiences in professional development, and interviews with teachers to illustrate how students' work can serve as a powerful professional development tool. The booklet provides staff developers with advice on arranging worthwhile professional development experiences for teachers.

Fennema, Elizabeth, Thomas P. Carpenter, Linda Levi, Megan Loef Franke, and Susan Empson. *Cognitively Guided Instruction: Professional Development in Primary Mathematics.* Madison, Wis.: Wisconsin Center for Education Research, 1990.*

This professional development program is intended to help elementary teachers develop a deep understanding of children's thinking and to then use that understanding to plan instruction. Cognitively Guided Instruction (CGI) is designed as a long-term (two- to four-year) program of professional development for mathematics teachers in the primary grades. Parts of the program are also effective at illustrating students' thinking to preservice or in-service teachers. CGI includes reading materials and seven videos for teacher participants and facilitation materials for teacher educators.

Ginsburg, Herbert P., Susan F. Jacobs, and Luz Stella Lopez. *The Teacher's Guide to Flexible Interviewing in the Classroom: Learning What Children Know about Mathematics.* Needham Heights, Mass.: Allyn & Bacon, 1998.

This book aims to help elementary school teachers understand and use flexible interviews to assess children's thinking about mathematics. The authors draw on the experiences of actual teachers to describe the process and results of implementing this form of authentic assessment in ordinary classrooms. The book closes with sample interview questions addressing major topics in elementary school mathematics.

Hiebert, James, Thomas P. Carpenter, Elizabeth Fennema, Karen C. Fuson, Diana Wearne, Hanlie Murray, Alwyn Olivier, and Piet Human. *Making Sense: Teaching and Learning Mathematics with Understanding.* Portsmouth, N.H.: Heinemann, 1997.*

This book is written by the principal investigators of four different research and development projects all centered on facilitating mathematical understanding and communication in elementary school classrooms. It comprises two parts—the first part describes features of classrooms that promote learning mathematics with understanding, and the second part illustrates these features in actual mathematics lessons. As a professional development resource, this book might stimulate teachers or potential teachers of elementary school mathematics to consider what it takes to create a culture of mathematical reflection and communication in their classrooms. The lessons depicted in the second part might also serve as cases to analyze and discuss.

Kamii, Constance. *A Teacher Uses Piaget's Theory.* Videotape series. New York: Teachers College Press, 1989.

This series of videos is intended to illustrate to practicing or prospective teachers what constructivist teaching and learning looks like in real primary-grade classrooms. The videos portray children creating their own meaning and strategies for mathematical operations, expressing their mathematical ideas to others, and questioning strategies that they do not understand. Titles in the series include "Double-Column Addition," and "First Graders Dividing 62 by 5."

Kenney, Patricia Ann, and Edward A. Silver. "Probing the Foundations of Algebra: Grade-4 Pattern Items in NAEP." *Teaching Children Mathematics* 3 (February 1997): 268–74.

Appearing in the "Algebraic Thinking Focus Issue" of Teaching Children Mathematics, *this article examines what elementary students seem to know and understand about patterns as indicated by students' responses to items on the 1992 Fourth-Grade NAEP. The tasks and students' work presented in the article might be useful in a professional development setting to generate discussion on students' thinking, on assessment, or on the development of algebraic ideas in the elementary grades.*

Lamon, Susan J. *Teaching Fractions and Ratios for Understanding: Essential Content Knowledge and Instructional Strategies for Teachers.* Mahwah, N.J.: Lawrence Erlbaum Associates, 1999.*

This book is intended to broaden the way that preservice or in-service teachers think about fractions and ratios. The author's notion is that once teachers understand fractions and ratios beyond traditional algorithms, they will plan instruction to foster learning with understanding in their own classrooms. The book includes samples of work from elementary and middle school students and is accompanied by a volume that provides extensive discussion of the tasks, the teaching problems, and other considerations raised throughout the book.

Miller, Barbara, and Ilene Kantror. *A Guide to Facilitating Cases in Education* and *A Casebook on School Reform.* Portsmouth, N.H.: Heineman, 1998.*

The guide offers an introduction to the use of case methods in teacher education and also provides essential information on effectively facilitating case materials and case discussions. The ideas and suggestions offered in the guide are illustrated through reference to an actual case. The casebook provides six ready-to-use cases with accompanying facilitation materials. The cases focus on school reform issues in mathematics and science such as assessment, changes in curriculum and instructional practices, and sustaining and extending reform efforts on school, district, and state levels.

Mokros, Jan, Susan Jo Russell, and Karen Economopoulos. *Beyond Arithmetic: Changing Mathematics in the Elementary Classroom.* Palo Alto, Calif.: Dale Seymour Publications, 1995.*

This book presents a rationale for reforming mathematics teaching in elementary school classrooms that is based on a constructivist philosophy of teaching and learning. The authors supplement their ideas with research support, practical advice, vignettes from actual classrooms, and examples of students' work and dialogue. The book also provides insight into the development and intended goals of the standards-based curriculum Investigations in Number, Data, and Space *by the authors.*

Moon, Jean. *Developing Judgment: Assessing Children's Work in Mathematics.* Portsmouth, N.H.: Heinemann, 1997.

The cases in this book depict elementary school teachers and their principals meeting in weekly study group sessions to assess children's work in mathematics classes. Each session addresses a major idea of assessment reform and how that idea might be implemented in the classroom. In addition to improving teachers' skills in assessing students' work, the author also provides a framework for creating a study group similar to the one featured in the book.

National Council of Supervisors of Mathematics (NCSM). *Great Tasks and More: A Source Book of Camera-Ready Resources on Mathematics Assessment.* Golden, Colo.: NCSM, 1996. (Available from NCSM, P.O. Box 10667, Golden, CO 80401; tel. 303-274-5932.)

This book is divided into four sections. Section 1 consists of ready-to-use blackline masters preorganized into a presentation that states the need for assessment reform and discusses Assessment Standards for School Mathematics *(NCTM 1995). Section 2 includes released performance assessment tasks for elementary, middle school, and high school levels and also examples of performance-based final exams. Section 3 contains samples of general rubrics, general student directions for performance assessments, and a guide to creating rubrics. Section 4 includes background readings in their entirety and references to other assessment resources.*

Richardson, Kathy. *Math Time: A Look at Children's Thinking.* Norman, Okla.: Educational Enrichment, 1990.*

This set includes two videos for viewing by participants and a video guide to assist the teacher educator. The videos, entitled Assessment Techniques: Beginning Number Concepts *and* Assessment Techniques: Number Combinations and Place Value, *model to teachers of primary-grade mathematics how to assess children's thinking about number concepts. The materials also provide sample responses from children at different levels of development and include handouts for teachers to use in assessing their own students.*

Richardson, Kathy. *Math Time: Thinking with Numbers.* Norman, Okla: Educational Enrichment, 1997.*

This set includes two videos for viewing by participants and a video guide to assist the teacher educator. Its purpose is to support teachers of primary-grade mathematics as they allow children to create their own strategies and understandings for number operations. The videos emphasize children's thinking in actual mathematics lessons and portray the same students over different grade levels to illustrate their mathematical progress. The video guide offers extensive support for facilitators, as well as activities and handouts for participants.

Romagnano, Lew. *Wrestling with Change: The Dilemmas of Teaching Real Mathematics.* Portsmouth, N.H.: Heinemann, 1994.*

This book describes the dilemmas encountered as the author and a teacher colleague collaborate to change the way mathematics is taught and learned in their ninth-grade general mathematics classrooms. The book is intended to stimulate teachers to analyze their own practice and consider the need for change. With its vignettes of actual classroom episodes, this book could generate discussion among preservice or in-service teachers on the process of change, classroom discourse, or teacher collaboration, at the same time motivating teachers to reflect on their own teaching.

Russell, Susan Jo, David Smith, Judy Storeygard, and Megan Murray. *Relearning to Teach Arithmetic.* Palo Alto, Calif.: Dale Seymour Publications, 1999.

This professional development program is divided into two packages, Addition and Subtraction *and* Multiplication and Division, *each of which includes three videos and a study guide for use with teacher participants. The videos portray real students working to develop their own strategies for whole-number operations, and the study guide helps teach-*

ers to analyze and discuss the mathematical thinking and understanding of the children portrayed in the videos. In its entirety, the program can provide up to ten two- to three-hour professional development sessions.

Schifter, Deborah, Virginia Bastable, and Susan Jo Russell. *Developing Mathematical Ideas: A Curriculum for Teacher Learning.* Palo Alto, Calif.: Dale Seymour Publications, 1999.*

This curriculum is intended to help elementary school teachers think deeply about the big ideas of elementary school mathematics. The curriculum integrates written and video cases, math activities, and reflective writing assignments into a coherent seminar for forty-eight hours of preservice or in-service education. Currently available are two packages, Number and Operations: Building a System of Tens, and Number and Operations: Making Meaning for Operations. Each package includes a casebook written by elementary school teachers depicting actual classroom episodes, a video portraying students engaged in mathematics lessons in real elementary school classrooms, and a facilitation guide to provide extensive support for teacher educators.

Schifter, Deborah, ed. *What's Happening in Math Class? Envisioning New Practices through Teacher Narratives.* Vol. 1. New York: Teachers College Press, 1996.*

Schifter, Deborah, ed. *What's Happening in Math Class? Reconstructing Professional Identities.* Vol. 2. New York: Teachers College Press.*

In these books, teachers provide their thoughts on the process of reforming mathematics instruction in their own classrooms. Each chapter includes several narratives, written by actual teachers, which collectively illustrate a principle of instructional reform recommended by NCTM. Each chapter closes with a commentary by a noted teacher educator that bonds the narratives together and stimulates teachers to look more deeply at how the featured principle is represented in each of the narratives. The narratives could serve as cases of exemplary standards-based practice for practicing or prospective teachers to analyze and discuss. Volume 1 features teachers of grades K–12, and Volume 2 includes only elementary school teachers.

Schoenfeld, Alan, Hugh Burkhard, Judah Schwartz, and Sandra Wilcox. *Balanced Assessment for the Mathematics Curriculum.* Palo Alto, Calif.: Dale Seymour Publications, 1999.

Balanced Assessment *offers assessment packages at the elementary, middle school, and high school levels. Each package consists of ten to twenty classroom-tested performance tasks that together encompass both the content and process strands of the NCTM Standards. For each task, the authors provide information on the main mathematical ideas, tell how to manage the assessment of the task, provide samples of typical students' responses at different levels of performance, and tell how to analyze the mathematics in students' responses. Each package could serve as an example of a performance assessment that meets the assessment Standards, or it could also provide teachers and teacher educators with a collection of rich mathematical tasks.*

Smith, Margaret Schwan, and Mary Kay Stein. "Selecting and Creating Mathematical Tasks: From Research to Practice." *Teaching Mathematics in the Middle School* 3, no. 5, (February 1998): 344–50.

Based on research conducted under the auspices of the QUASAR project, this article presents a model for categorizing mathematical tasks according to levels of cognitive demand. The authors describe a task-sort activity and task analysis guide they have used in professional development workshops to help teachers identify mathematical tasks having the potential to elicit high levels of cognitive thinking and reasoning. The article includes the task analysis guide, samples of tasks used in the task-sort activity, and examples of specific tasks at each level of cognitive demand.

Stein, Mary Kay, Margaret S. Smith, Marjorie A. Henningsen, and Edward A. Silver. *Implementing Standards-Based Mathematics Instruction: A Casebook for Teacher Professional Development.* New York: Teachers College Press, 2000.

At the heart of this book are six instructional cases, each of which takes you into the classroom of an urban middle school teacher who is attempting to enact reform-oriented mathematics instruction. The book discusses how these cases are situated into a larger framework of ideas on mathematics teaching and learning that is based on analyzing cognitively challenging mathematical tasks and how such tasks unfold in actual classroom lessons. The mathematical content of the cases includes (1) fractions, decimals, and percents; (2) multiplying fractions; (3) mean, median, mode, and range; (4) multiplication of monomials and binomials; (5) data organization and analysis; and (6) problem solving.

Trafton, Paul R., and Diane Thiessen. *Learning through Problems: Number Sense and Computational Strategies–A Resource for Primary Teachers.* Portsmouth, N.H.: 1999.*

This book is a resource to help teachers of primary-grade mathematics to understand and implement problem-centered instruction and learning in their classrooms. The book describes the idea of problem-centered instruction, applies this idea to the teaching of number sense and computational strategies in grades K–2, and discusses how teachers would go about implementing this approach in their classrooms. Examples of students' work and references to national standards illustrate and support the authors' ideas.

U.S. Department of Education, Office of Educational Research and Improvement (OERI). *Attaining Excellence: A TIMSS Resource Kit*. Washington, D.C.: U.S. Department of Education, OERI, 1997.*

This kit consists of four modules entitled TIMSS as a Starting Point to Examine: (1) U.S. Education, (2) Student Achievement, (3) Teaching, and (4) Curriculum. *Of particular interest is the teaching module, which features an eighty-minute VHS tape of eighth-grade mathematics lessons thought to typify instruction in the United States, Japan, and Germany. These lessons provide an international perspective through which teachers can examine their own practice. The tape is accompanied by a thorough moderator's guide that includes presentation handouts and transparencies, advice on leading the discussion, references to related research, answers to FAQs, and lesson transcripts.*

WGBH Boston. *Teaching Math: A Video Library, K–4*. South Burlington, Vt.: The Annenberg/CPB Math and Science Collection, 1995.*

This series consists of (1) forty-five twenty-minute video segments of mathematics lessons taken from actual elementary school classrooms and (2) a guidebook providing teacher educators with suggestions on facilitating discussions and activities using these videos. The lessons demonstrate real elementary school teachers' efforts to teach in ways that are consistent with recommendations from NCTM. They cover a breadth of content areas while emphasizing problem solving and communication. The videos were filmed in a variety of schools across the country: urban, suburban, rural, small, large, and bilingual.

WGBH Boston. *Teaching Math: A Video Library, 5–8*. South Burlington, Vt.: The Annenberg/CPB Math and Science Collection, 1997.*

This series consists of (1) three videotapes, each containing two fifteen-minute segments of mathematics lessons taken from actual middle school classrooms and (2) a guidebook providing teacher educators with suggestions on facilitating discussions and activities using these videos. The lessons depict real middle school teachers attempting to teach in ways that are consistent with recommendations from NCTM. They cover a breadth of content areas while emphasizing problem solving and communication.

WGBH Boston. *Teaching Math: A Video Library, 9–12*. South Burlington, Vt.: The Annenberg/CPB Math and Science Collection, 1996.*

This series consists of (1) fifteen 15-minute segments of mathematics lessons taken from actual high school classrooms and (2) a guidebook providing teacher educators with suggestions on facilitating discus-

sions and activities using these videos. The lessons depict real high school teachers attempting to teach in ways that are consistent with recommendations from NCTM. The videos focus on reasoning and problem solving and feature a range of mathematical ideas.

Wilcox, Sandra K., and Perry Lanier. *Using Assessment to Reshape Mathematics Teaching.* Hillsdale, N.J.: Lawrence Erlbaum Associates, 1999.*

This casebook consists of seven decision-making cases in ready-to-use format and their accompanying facilitation materials. Each case pushes mathematics teachers to use multiple sources of assessment, such as listening to students and analyzing students' work, to determine what students understand, how they understand it, and what this implies for the teacher's next instructional moves. The casebook is supplemented by a sixty-minute video with clips for four of the cases. Five of the cases are from middle-grades classrooms, one is from grades 3–4, and one is from high school.